DETAILING TECHNIQUES

Make Your Car Look Its Best

David H. Jacobs, Jr.
Photography by Colin Date

The Complete Street Machine Library®

Detailing Techniques

Make Your Car Look Its Best
Printed in 2006.

Published by National Street Machine Club under license from MBI Publishing Company.

Tom Carpenter
Creative Director

Heather Koshiol
Managing Editor

Teresa Marrone
Book Design and Production

Jen Weaverling
Senior Book Development Coordinator

1 2 3 4 5 6 / 10 09 08 07 06
ISBN 1-58159-269-8

© David H. Jacobs, Jr. and Colin Date, 2003, 2005.

All rights reserved. No part of this publication may be reproduced, stored in an electronic retrieval system or transmitted in any form or by any means (electronic, mechanical, photocopy, recording or otherwise) without the prior written permission of the copyright owner.

National Street Machine Club
12301 Whitewater Drive
Minnetonka, MN 55343
www.streetmachineclub.com

About the Authors

David H. Jacobs, Jr. has been busy writing books about different automotive topics since the 1980s. He likes to talk to and hang around with experts in the field to learn how they best accomplish their work in automotive detailing, bodywork, painting, and the like. He appreciates and especially enjoys writing about their tips, advice, and tricks of the trade.

Photographer **Colin Date** has written and photographed numerous articles for magazines such as Chevy High Performance, Mopar Muscle, Muscle Car Review, and Mustang Monthly. Colin and his family reside in Southern California.

Author's Acknowledgments

I would like to thank some folks for helping me put this book project together. At the top of the list is Colin Date. He shot the pictures for the book and did a great job. I would envision a project and jot down a note or two, and he would just take off and make it happen. We both must also thank Lenny Amato and his crew at Amato's Detailing in Laguna Beach, California, for their help with photos. Their time and infinite patience allowed Colin to shoot a lot of pictures and get some "insider" detailing tips.

Next is my buddy Art Wentworth. He thoroughly enjoys automobiles and actually likes spending time to detail them. His knowledge and experience as an avid auto enthusiast are second to none.

John Sloan from the Eastwood Company and the folks at Meguiar's, Mothers, Simple Green, Dent Fix Corporation, System One, and Pro Motorcar Products were all most helpful in providing information about their cosmetic car care products, tools, and materials. I very much appreciate their support.

My wife, Janna, and son, Luke, also took time out of their busy schedules to help out any way they could. From getting copies of stuff made to helping with detailing chores, they were always up for the challenge. I appreciate their patience in putting up with me, too.

Finally, I want to thank my editor, Peter Bodensteiner, and the folks at MBI Publishing. They are a pleasure to work with and are always ready to help out at a moment's notice. I really appreciate the way they make the material I send them look and read so much better.

Contents

Introduction 4
Understanding the Information Boxes 5

Chapter 1—Getting Ready — 6

Chapter 2—Preliminary Washing — 14
Technique 1: Basic Vehicle Wash 19
Technique 2: Doorjambs, Hood, and Trunk Edges ... 22
Technique 3: Removing Tree Sap, Bug Splatter, and the Like 24
Technique 4: Grille, Brightwork, and Running Boards ... 26
Technique 5: Tires and Wheels 28
Technique 6: Vinyl Tops and Convertibles 30
Technique 7: Removing Stickers and Decals 32
Technique 8: Underbody 34

Chapter 3—Interior Detailing — 36
Technique 9: Vacuuming 39
Technique 10: Cleaning the Interior and Dash 41
Technique 11: Doorjambs and Panels 44
Technique 12: Vinyl and Rubber Dressing 47
Technique 13: Vinyl Seats 49
Technique 14: Cloth Seats 51
Technique 15: Leather Seats 53
Technique 16: Shampooing Carpets 55
Technique 17: Cloth Protectants 57
Technique 18: Interior Extras 58

Chapter 4—Under the Hood — 60
Technique 19: Initial Cleaning 64
Technique 20: Initial Cleaning without Water 67
Technique 21: Detailed Cleaning 69
Technique 22: Painting the Engine Block and
              Other Engine Parts 72
Technique 23: Polish and Shine 75
Technique 24: Extras 76

Chapter 5—Exterior Shine — 78
Technique 25: Removing Severe Oxidation 82
Technique 26: Clay Bar 84
Technique 27: Buffing with a Machine 86
Technique 28: Hand Polish 89
Technique 29: Hand Wax 92
Technique 30: Removing Polish and Wax Residue 94
Technique 31: Paint Blemishes and Chips 96

Chapter 6—Underbody Detailing — 98
Technique 32: Steam Cleaning 101
Technique 33: Thorough Rinsing and Cleaning 102
Technique 34: Fenderwell Painting or Undercoating . 104
Technique 35: Frame Members and Tailpipes 106

Chapter 7—Glass, Trim & Moldings — 108
Technique 36: Removing Stickers and Decals
              from Glass 111
Technique 37: Glass Cleaning 112
Technique 38: Glass Polishing 114
Technique 39: Removing Buildup from Window Edges . 115
Technique 40: Trim Cleaning 116
Technique 41: Chrome, Rubber, and Vinyl 118
Technique 42: Moldings 120
Technique 43: Windshield Repair 121
Technique 44: Plastic Window Care 122
Technique 45: All the Little Extras 123

Chapter 8—Tires & Wheels — 126
Technique 46: Whitewalls and Raised White Lettering .. 129
Technique 47: Dressing and Tire Black 131
Technique 48: Chemical Wheel Cleaners 133
Technique 49: Wire Wheels 134
Technique 50: Mags and Other Special Wheels 136
Technique 51: Wheel Covers and Hubcaps 138
Technique 52: Painting Wheels 139

Chapter 9—Finishing Touches — 140
Technique 53: Trunk Rejuvenation 143
Technique 54: Parts Replacement 146
Technique 55: Just Nice or New Car Smell 148
Technique 56: Bra Care and Car Covers 149
Technique 57: Concours d'Elegance Considerations . 151
Technique 58: Final Inspection 154
Technique 59: Paintless Dent Removal 157

Index 159

Introduction

Automobile detailing is a systematic process intended to make every part of a vehicle looks its best. From the top down, this classic looks crisp and stands tall. Paint is polished to perfection, wheels and tires look great, chrome sparkles, windows are clean, and fenderwells are tidy.

Automobile detailing is the process of making an automobile look its best without major dismantling or extensive repainting. Much more than a quick wash and vacuum, detailing requires planning and attention to the small stuff. Exterior detailing is a systematic cleaning, polishing, and waxing from front to back, side to side, and top to bottom. Interior detailing is also a systematic, labor-intensive cleanup that could take you right down to the floorboards.

Detailing might be described as cosmetic car care that falls somewhere between a run through the local car wash and total restoration. You can give your car a minimal detail on a Saturday afternoon with an exterior wash and an interior vacuum and wipe. A moderate detail will generally take an entire day and consist of a thorough exterior wash, polish, and wax, and a conscientious cleaning and dressing effort for the interior. An intense detail will require a number of days to clean, polish, wax, and dress every part of the vehicle, including the engine compartment, to near perfection. The final and highest degree of detailing is preparation for a Concours d'Elegance event, which can literally take months to complete.

Most used cars on a dealer's lot have been detailed. They exhibit shiny black tires, velvety paint, clean windows, fresh upholstery, and a glossy engine compartment; there is no question that these cars have been professionally, albeit generically, prepared.

One must realize that most dealers' used cars are detailed in just a few hours. In most cases, professionals must take shortcuts to get cars looking their best in the least amount of time. They do this to keep dealers happy and keep labor costs at a minimum.

Due to these constraints, generic detailers commonly rely on clear lacquer paint to make engine compartments, trunks, doorjambs, and vents look shiny and new. They use ample amounts of dressing on interiors to quickly achieve the best cosmetic appearance in the shortest time. Paint is buffed and then hand waxed with a liquid for speed and high gloss.

These cars look good, but they also look "detailed." They look like something that just came out of a gloss factory. Everything is too glossy and too shiny, and the cars don't look real, especially when they are compared to identical vehicles that have been meticulously detailed by hand.

Harsh chemicals clean quickly, and they are safe when used properly. Misuse can result in paint blemishes, fabric stains, and other damage. That's why most car enthusiasts prefer old-fashioned soap, water, and lots of elbow grease. Professionals cannot always afford the time it takes to detail cars without the use of chemicals. But when they do detail by hand, using very mild methods and taking days to gently bring automobiles to their fullest potential, the pampering may cost $500 or more.

Serious car people regard detailing as a two-step process. First is cleaning, polishing, and waxing. Second is getting things right, like license plate frames, floor mats,

Below: A full-blown restoration was needed to make this special automobile look better than new. Take note of the perfect paint finish, with no hint of swirls or polish/wax buildup anywhere. Everything is just right; not too much gloss, just clean, tidy, and crisp.

Below Bottom: The engine compartment of your special vehicle can be made to look like this. All it takes is a little time and attention to detail. Every part under the hood was thoroughly cleaned and then painted, polished, or dressed. The key is to get things clean before you start shining.

Introduction

Below: Automobile interiors deserve the efforts you put forth to make them look this nice. This is, after all, the place where you and your friends will sit to enjoy casual rides and road trips. It is a pleasure to drive a clean car—it even seems to run better when it looks this good.

Below Bottom: No matter what type of vehicle you drive, you can make it look good with time, patience, and elbow grease. It took a while for your car or truck to get dirty and lose its shine, so don't expect to get it looking new in just a few hours. Put together a plan, and then follow along as the projects in this book help you detail your automobile to perfection.

and jacking equipment. Does the license plate frame blend with the back of the vehicle? Simple chrome frames look good on cars with chrome bumpers. Black frames look best on vehicles with black bumpers and those with black paint around the license plate area. When you put new tabs on the license plate, take pains to place them on square. This is a minor consideration, but one that will catch your eye. You can also line up the screws holding the license plate in place.

Some cars come equipped with tool kits. Look at them once in a while to make sure all tools are in working order. Take a few minutes to lubricate the pliers and rustproof the rest of the tools with a light coat of WD-40. The jacking equipment is equally important.

Decals and stickers are more popular now than ever before. Some, like parking permits, are necessary. Instead of placing them just anywhere, take a minute to find an appropriate spot. If a window seems most logical, place the sticker in the corner and be sure the bottom edge is square with the window trim. If the bumper is more suitable, keep the sticker square with the bumper's bottom edge. Never put a sticker on the painted surface of the trunk. If you ever attempt to remove it, you stand a good chance of taking paint with it. Once the sticker is removed, you'll notice the paint underneath is darker than the paint on the rest of the car.

The Bottom Line

The ultimate secret to detailing is time, patience, and elbow grease. The more quality time you spend, the better results you will achieve. If your time is limited, spend it on one area, as opposed to a little time on a lot of areas. Take one day to clean nothing but the wheels. Use another day on the interior. Plan a weekend for the exterior and another for the engine.

You should also realize that a complete and thorough detail is only necessary once, providing you have a consistent system for regular maintenance. Car enthusiasts routinely go through their rigs once or twice a year. Then they wash their vehicles weekly, waxing one section each time. They concentrate on the hood, top, and trunk in the summer, sides in the winter. It doesn't take too much time, and it ensures adequate wax protection.

Keep detailing as simple as possible. Dismantle as necessary to reach dirt around parts such as the grille, moldings, emblems, seats, and knobs. After that, stay on top of those areas by cleaning frequently, not giving dirt a chance to build up. If you question the use of certain chemicals, always remember that you won't go wrong using mild soap, water, and a soft touch.

A perfect detail followed by regular weekly and monthly maintenance is the best way to keep an automobile looking good, standing tall, and remaining valuable.

Understanding the Information Boxes

The example below shows how each information box is set up at the beginning of each project and the information it contains.

TECHNIQUE X INFORMATION BOX

Time: The amount of time it should take to complete the project

Tools: The tools and supplies you will need to do the job

Talent: The skill needed to do the work; 1 = ★, a novice, and 5 = ★★★★★, an expert

Tab: The estimated cost of the project

Tip: Suggestions on how to best complete a project or what else to look for

Gain: Why you should do the project

Complementary project: A project that would be timely or practical to do at the same time

Chapter 1
Getting Ready

To do a great detailing job, you need to know which products, tools and accessories are essential.

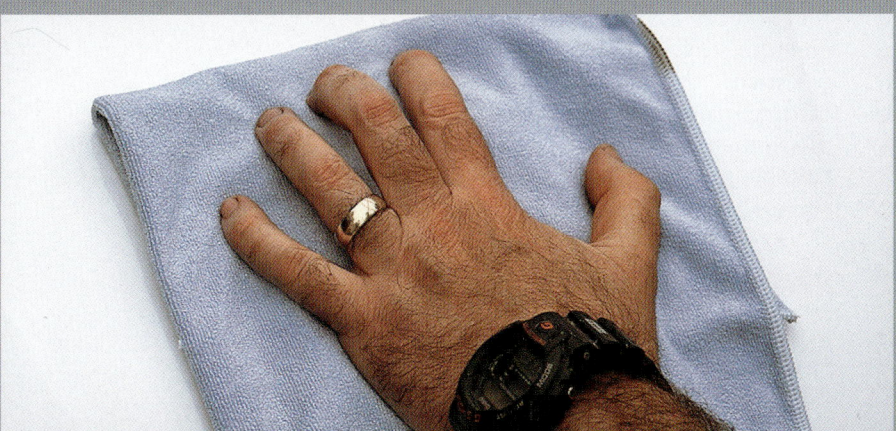

Work Area, Tools, Materials, and Supplies

Many detailers and automotive enthusiasts have successfully used the tools, materials, and supplies featured in the following projects. It is imperative that you follow the instructions on the label of each product you use. If you question directed use, seek advice from an autobody paint and supply store or your local professional detailer.

Appropriate Attire

During the course of detailing, your hands, arms, and body will come in contact with the vehicle. If you wear rings, imagine what will happen if your hand slides off a towel and onto the painted surface. Watches and bracelets can also scratch paint. Remove all jewelry before detailing.

Clothes are important. You want to be comfortable while bending and twisting to reach all parts of the automobile. Be wary of jeans that have rivets at the pockets. You will be leaning over the fenders to reach parts in the engine compartment. A cotton warm-up suit is great. It allows you easy and comfortable movement and has no protruding rivets or zippers that can cause scratches.

Tennis shoes are fine as long as they don't have treaded soles that can trap dirt and debris. When you sit in the interior to clean, you won't want all that stuff coming off your shoes and onto the carpet. If you must wear shoes with such soles, lay a towel on the floor for your feet.

The work area you select to wash your car or truck should have a slight downhill slope. This will help water to run off away from the vehicle, eliminating pesky puddle problems. Use caution if you clean your car with a pressure washer! A low-pressure garden hose is preferred.

Some detailers wear a long apron while buffing and waxing. The apron serves two purposes. First, it keeps the detailer's clothes clean. Second, it protects the body of the vehicle while the detailer has to lean against it.

Work Area

The area in which you choose to work should be appropriate for the type of job that you are planning. For a preliminary cleaning that includes an engine compartment, you must realize the amount of dirt, grease, and grime that will fall off onto the ground. Is your driveway suitable for this mess, or should you find somewhere else?

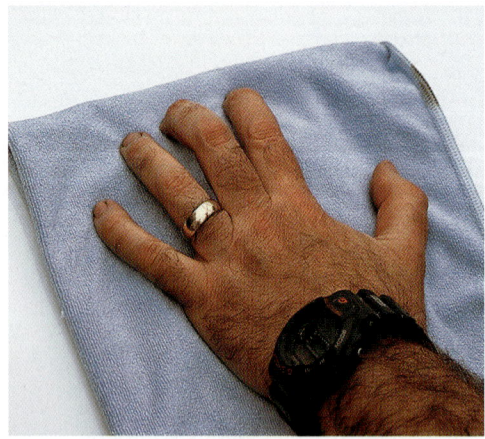

Far Left: The very first rule in automotive detailing is to remove your watch, rings, and anything else from your hands and arms that could scratch paint or other surfaces. This rule must not be broken if you are serious about making your ride look its best.

Left: Comfortable clothes will help to make your detailing endeavor more enjoyable. In addition, you must be aware that metal rivets, zippers, and other hard objects on clothing will pose scratch hazards to your vehicle's paint finish.

Getting Ready

If you plan to clean your vehicle's dirty engine compartment using solvent and water spray, consider doing it at a self-serve car wash. Most of these facilities feature special drains and runoff systems that trap dirt and grease before they get into the common municipal drainage system.

Good places to wash engines and fenderwells are self-serve car washes. They offer high-pressure water and soap mixed through a high-pressure wand. Each bay is equipped with drains in the floor, allowing for heavy residue runoff. Self-serve car washes are inexpensive and useful places where you can work to remove heavy accumulations of dirt and grease. Afterward, relocate to your driveway for more detailed cleaning.

Use caution with the high-pressure wash. Aiming the tip of a wand too close to stickers in the engine compartment will remove them. Some paint will come off the engine, which is to be expected. You can repaint later. Watch out for water ricochet. Pointing the wand at the top of the intake manifold, with your head directly behind the wand, may splatter water and grease in your face and possibly your eyes. Stand to the side and plan for splatter to go away from you; consider wearing safety goggles.

It is not a good idea to use high-pressure water on vehicle bodies, except for rinsing during engine cleaning. The force of the water could damage paint and emblems. Save the body wash for your driveway.

You can do all of the preliminary washing in your own driveway or yard if the car is reasonably clean. Just remember that car washing should be done in the shade, although you can do it in the sun as long as you rinse frequently. The object is to prevent soap from drying on the car. Wash the shady side first, sunny side last. Dry the entire car immediately after washing.

Almost all work can be done on the driveway. You can clean the trunk, paint the engine, and detail the interior. An exception may be waxing, which must be done in the shade. A garage or carport is perfect. The sun dries wax too fast, which makes it hard to remove and results in streaks. If one of your favorite hobbies is car waxing at the beach or park, find a shady spot. If nothing else, wax the shady side first; then reposition the vehicle so the opposite side is in the shade.

Washing Equipment
Water Applicators

Portable pressure washers offer the same high water pressure as self-serve car washes for cleaning engines and underbodies. They can be rented at any local tool rental yard. Your need for one is disputable; other than for washing the engine and underbody on a neglected vehicle, you shouldn't need one, and it is far cheaper to use the self-serve car wash. High-pressure water sprays can help to clean wheels, grilles and vinyl tops, but you must consider the damage that can be done when they are not used properly. Decals and stickers can be blown off, paint peeled, and vinyl torn. In the long run, it is much safer to use the gentlest techniques available.

Ordinary household water pressure is sufficient for most purposes. If higher pressure is needed, use a high-pressure nozzle. Be careful not to drop such a nozzle in a way that it could careen off the ground and hit your car.

An open-ended garden hose is best for car washing. Softly flowing water will remove soap and leave the waxed paint almost dry because water will fall off the car in sheets. Applying water through a nozzle will result in thousands of small water puddles, an added drying chore.

Most auto parts stores carry multiuse wash wands. They are long metal or plastic tubes with a hose connection at one end and a brush at the other. Soap is mixed with water and comes out through the brush. Although they are advertised as work savers, one must wonder what effect they have on automotive paint.

Wands like these may be well suited for washing windows, vinyl house siding, and garden furniture, but they are not for quality cars. The brush will leave fine scratches. Scrubbing a particularly tough spot will result in deeper scratches. It may be all right for the person who has a regular driver that gets washed once every six months, but not the person who cares enough about his or her car to wash it every week.

Car Wash Soaps

Almost every serious auto enthusiast has a different preference for car wash soap. Some feel mild dish soaps are best; others prefer name-brand products

Plan to use a quality car wash soap to clean the outside of your automobile. Although some folks have used regular dish soaps for years with little trouble, you must realize that companies like Meguiar's and Mothers have spent a lot of time putting together products that do a good job of cleaning and go easy on waxed finishes.

Detailing Techniques

Far Left: Invest in a quality cotton chenille wash mitt. These mitts are soft and will not cause swirls or spider webbing to occur on your car's paint finish. Mitts are easy to handle and also protect the skin on your hands while cleaning grilles and other areas with sharp edges.

Left: As important as soft cotton wash mitts are the towels or cloths you expect to wipe on your vehicle's surface. Soft cotton terry towels work well to help dry cars after a wash and also to remove dry polish and wax during paint care projects.

designed specifically for washing automobiles.

Most agree, however, that powdered soaps are out of the question. They feel that a single grain of powder not dissolved completely can scratch paint. Shy away from household cleaners for car washing, too, as they can be much too abrasive or alkaline. Save them for more intense cleaning chores on other parts of the vehicle.

The focus on which soap or detergent to use for washing car bodies seems to be which one has the least impact on removing car wax. Most dish soaps contain degreasers. While the degreasing feature is good for removing road grime, it could also remove car wax. This poses no problem if you always wax your car after each wash. On the plus side, some dish soaps might help to reduce water spotting.

Meguiar's, Mothers, and many other manufacturers of car care products make car wash soaps that have been tested on automobiles, not dishes. You can't go wrong using any of these products. Try each at different times until you find one that works best for you.

Wash Mitts

Washing cars is easiest with a soft cotton wash mitt. It holds plenty of soap and water, covers a good-sized area, and protects your hand. A few different types are on the market, including some made of synthetic materials that look like lamb's wool. Most car enthusiasts feel the synthetics are too rough. The myriad minute scratches left on paint after washing with a nylon mitt look like webbing left behind by 1,000 spiders. The thin lines are accented by sunlight and are exceptionally visible. To remove "spider webbing," polish with a sealer and glaze like Meguiar's Number 7.

Serious auto enthusiasts prefer soft cotton wash mitts, which are often made of chenille, a type of soft cord. The fibers are soft and do not promote swirls or spider webbing. Try using two cotton wash mitts, one for the vehicle body and one for the fenderwells, underbody, and engine compartment. A wash mitt will collect grit. It is very important to rinse it frequently. You should also rinse the wash bucket occasionally and refill with clean soapy water, remembering that anything collected in the mitt or the wash bucket is a potential scratch hazard.

You can also use towels for vehicle washing. They are cumbersome if they're too big, so cut them in half. Baby diapers are soft and a good size. The disadvantage of using cloths and towels is that your hand can slip off of them, resulting in cuts from sharp edges on emblems or other projections. Cloths and towels don't seem to hold as much soap as a mitt, either.

Brushes

To detail an automobile thoroughly, you need a small assortment of brushes. A 3-inch, soft, natural-bristled paintbrush works wonders around headlights, mirrors, window trim, emblems, and wheels. The soft bristles will not scratch paint and will reach areas your mitt or cloth cannot. Since the brush is soft, you may have to use more than one application on tough spots. Be patient. The value of using a nonscratching brush is well worth the effort. As a precaution,

Right: An old soft toothbrush does a good job of removing dried polish or wax residue from the groove between a chrome molding and its rubber base. Toothbrushes are used in all vehicle areas to help clean away dirt and debris from tight spaces.

Getting Ready **11**

  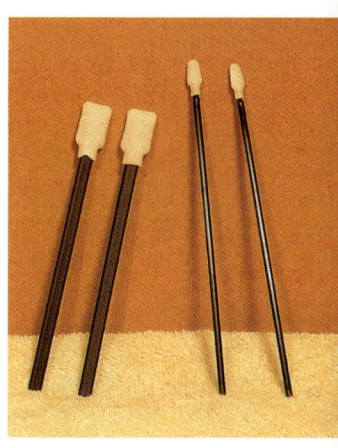

Above Left: This is an assortment of brushes found at an auto parts store. As you begin your auto detailing endeavors, you will discover a multitude of uses for brushes of all types. At the least, you'll need a stout scrub brush and a soft paintbrush for most projects.

Above Middle: A soft, floppy paintbrush works wonders when you are trying to clean areas such as gas filler spaces. Plan to use a paintbrush while washing the car to loosen up dirt and grime from around window trim, headlight bands, grooves, lips, edges, and other tight spaces.

Above Right: These are reusable swabs from Meguiar's. Available in two different sizes, they work great for getting into tight spots to remove dust and dirt. Use them on dashboard vents, between slim grooves on consoles, and in a host of other recesses that are generally tough to get to.

place a strip of duct tape around the metal frame of the brush. This will prevent paint damage while brushing in tight areas.

A smaller paintbrush, about 1 inch wide, works very well in close quarters. The small size allows easy access to spots around the grille, around lug nuts on wheels, and in tight pockets near taillights and bumpers.

A plastic-bristled scrub brush can be used on tires, fenderwells, and underbody areas. The stout bristles knock off dirt and debris and scrub rubber clean. This brush can also be used to shampoo carpets, upholstery, vinyl tops, and convertibles. If necessary, it can be used under the hood to remove extra-heavy accumulations of dirt.

Whitewall brushes are small, short-bristled wire brushes. They are not nearly as strong as the larger painter's wire brush. The bristles are closely cropped together and make short work of removing dirt and stains from whitewalls.

Toothbrushes are used in every detail shop. Old toothbrushes with soft bristles work best. These handy tools can be used to remove dirt and wax buildup around light lenses and in emblems, dirt in upholstery seams, stubborn spots on carpet, and almost anything else you deem necessary. Have more than one on hand; you may use one in the engine compartment, and you won't want to use it later on the interior.

Cotton swabs are also very useful. Their size, texture, and absorbing qualities are perfect for removing dust around radio buttons, cracks in dash panels, and slots around heater controls. They can also be used to apply and remove wax inside painted vent openings and around emblems.

Finally, a small 1- to 2-inch paintbrush, with the bristles cut to about 3/4 inch, works exceptionally well to remove wax from emblems, cracks, and trim. The bristles will not scratch, so the brush can be used as a final detailing tool. The short length of the bristles gives them the strength to remove wax buildup.

Steel Wool Soap Pads

SOS pads are a familiar item at most supermarkets. They are pads of steel wool impregnated with soap. Designed for scrubbing pots and pans, they also work well on chrome, whitewalls, vinyl tops, convertible tops, some wheels, and engines.

The texture of the pad is soft and should not scratch real chrome. The soap adds cleaning and degreasing power. Since they fall apart rather quickly, you will need more than one during the course of a complete detail.

Steel wool can be used for some cleaning jobs. Used with a window cleaner, it can remove stubborn bug residue from windshields. Used with wax, it can polish, clean, and protect real chrome in one application. Always use soft steel wool, like number 0000.

All-Purpose Cleaners

Detail shops use heavy-duty cleaners supplied by wholesale outlets. Some similar products are available at retail. Simple Green is a heavy-duty cleaner

12 Detailing Techniques

available at most auto parts stores. Many detailers have had good results using it for many cleaning jobs, including engines, tires, trunks, and wheels. It is a mild cleaner that penetrates dirt—just what you want. Other household cleaners may work just as well.

Use all-purpose cleaners to clean the engine compartment, jacking equipment, underbody, and vinyl. Such cleaning agents are too harsh for washing the painted surfaces on car bodies, but they work well for cleaning other parts. You may find the best results purchasing liquid cleaner in the gallon size, diluting it according to label instructions, and applying through squirt bottles. Simple Green and other cleaners also come in spray bottles that you can refill from the larger container.

Use this type of cleaner on vinyl interior parts. Spray it on one side of a folded towel to clean the headliner; use the other side of the towel to wipe dry. Do the same for dashboards, seats, and seatbelts. Clean ashtrays with this product and a toothbrush.

The brand you purchase is up to you. As in choosing car wash soaps, you have to find the cleaner that works best for you and your needs. Under no circumstances should you mix cleaners. They may react chemically and damage the material you are cleaning.

Toothpaste is a mild abrasive that can be used to clean stained metal parts. Some detailers have used it to scrub metal emblems. Since it has not been designed for use on cars, you may have better luck using chrome polish first, saving toothpaste for a last resort.

Upholstery and Carpet Cleaning Equipment
Shampoos

To clean carpet and upholstery completely, you need to use shampoo. Most supermarkets carry an assortment of such products. Used with a stiff plastic-bristled brush, shampoo makes most upholstery turn out beautiful. Use it as you would for household furniture. It is best to have a wet/dry vacuum cleaner available to remove excess water and soap. If not, use lots of dry towels to soak up moisture.

Carpets can be shampooed just like the upholstery. Fill a small bucket and spray bottle with shampoo mixed with water according to the label instructions. Use the plastic-bristled brush to work up a good lather. The spray bottle comes in handy for stubborn stains.

If carpeting is merely spotted, you should not have to shampoo the entire area. Supermarkets also carry many carpet cleaners designed for spot removal. These products do not require the use of water or a wet/dry vacuum cleaner. They are easy to use and work well.

Vacuum Cleaners

A wet/dry vacuum cleaner is most versatile. Generally stronger than household units, it picks up larger debris, as well as water. Unless you plan to shampoo the carpet and upholstery, there is no need to purchase or rent one. Dry carpet cleaners work well and the powdery residue can be removed with a dry vacuum cleaner.

If the interior of your car warrants the power of a wet/dry vacuum cleaner and you don't have one, drive to a local self-serve car wash. Most of these facilities have three or four vacuum units on site. They have wet/dry capabilities and can also be used to remove shampoo.

The small 12-volt vacuum cleaners powered by cigarette lighters are fine. Since they lack power, you can sweep debris into a pile with a plastic-bristled brush and then vacuum it. These units

Above Left: You'll need a strong cleaner to remove dirt and grime from engine compartments, interiors, and other automobile features and spaces. Simple Green works very well. It is most economical to purchase cleaner in a big container and fill squirt bottles as they become empty.

Above Right: Automotive carpet and upholstery cleaners are available in auto parts stores, some supermarkets, and variety stores. You can choose from two basic types—one that goes on wet and dries to be vacuumed up with a regular vacuum cleaner, or the wet shampoo, which requires use of a wet/dry vacuum for cleanup.

Getting Ready 13

Above Left: A vacuum cleaner is a must for cleaning auto interiors and trunks. Wet/dry vacuums are generally the most powerful and are used to help dry upholstery and carpet that has been wet shampooed. The crevice tool attachment works very well for most projects, while a soft brush attachment does a good job of cleaning out dashboard vents and the like.

Above Right: The System One advanced polishing system made the driver's side of this hood reflect like a mirror. The Pro-Kit comes complete with three buffing pads, sponge, microfiber towel, wax, and a special polish that is the heart of the system. Designed to remove wet-sanding scratches, this polish is used with all three pads to result in a perfect swirl-free finish. The System One Pro-Kit seen here can be purchased through the Eastwood Company for about $70.

do not have enough power to loosen stubborn grit, hence the need for a brush. The same holds true for battery-powered vacuum cleaners.

Drying Cloths

You will need towels to dry the car and clean areas other than the body. Bath towels are great, except for their size. Cut them in half and fold into quarters. Medium-sized fluffy hand towels with long naps are perfect. Nap refers to the surface texture of cloth, towel, or carpet. Unlike flannel, which is smooth, towels and carpet are made with thick woven fibers (pile) that stick up out of the base material and form a soft cushiony texture.

Drying the car requires at least two towels. If you used free-flowing water, rather than a nozzle, there won't be much water on the car. Use the first towel to absorb most of the water and the second to dry the car completely. Later, use the damper towel to wipe doorjambs, trunk lid, and hood edges. Use the other damp towel to wipe off light dust and dirt from the dashboard, gauges, and other interior surfaces.

You will need towels or cloths for other cleaning chores, such as sprucing up the interior. The coarse nap on hand towels helps remove dirt and embedded grime. Folding them in quarters allows you to refold and use clean sections as needed.

Paper towels can be used to wipe greasy parts and plug wires. Inexpensive and plentiful, they are useful for the dirtiest and greasiest jobs. With their absorbing qualities, paper towels can also be used to keep ignition parts and the carburetor dry during engine cleaning.

Wax application and removal require a different set of cloths. Some waxes come with a round, spongelike applicator. These are OK, but you may have better luck using a rectangular sponge. The straight edges on the sponge allow greater control while waxing along seams and sunroof edges. The sponge may be cleaned and reused. Since they are quite inexpensive, having an ample supply on hand is easy.

Removing wax raises another point of disagreement among auto enthusiasts. What cloth works best for rubbing and polishing? The only criterion is that the material be very soft, inexpensive, and easy to use. Again, fold the cloth into quarters and refold as a side becomes dirty.

Baby diapers are a favorite. They have never been exposed to harsh materials that may scratch, such as metal chips or grit. The only problem may be availability. You can buy new ones, but you must consider the cost. Cheesecloth is another option that works well. The only drawback is that it can only be used once. Old cotton shirts are fine, but how many do you have hanging around? Another option may be the fabric store. Some enthusiasts purchase soft white flannel at the fabric store and just pop the cloths into the clothes washer and dryer after use. Flannel actually seems to get softer the more it is washed. As cloths wear out, purchase another yard or two for just a few dollars.

Chapter 2
Preliminary Washing

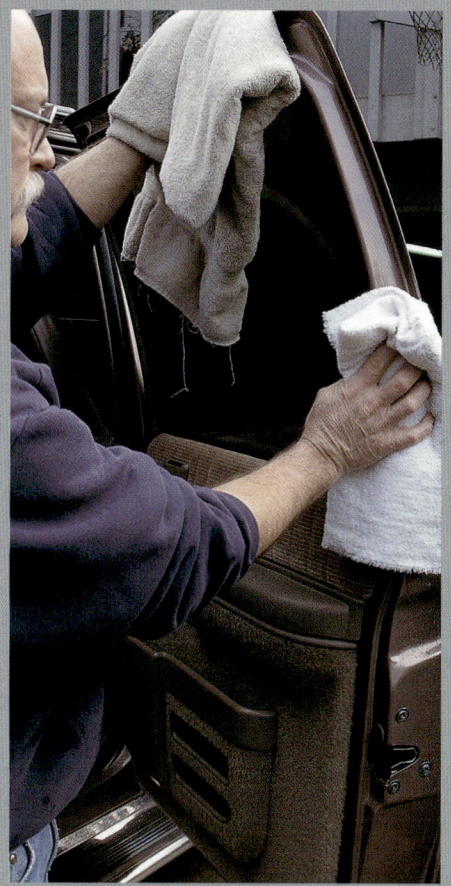

Some people think an external wash is the whole show. But it's really just the start. And it takes a lot more than a bucket of suds and a hose to do the job right.

Preliminary washing is the first step in a thorough detail job. If your automobile has been reasonably maintained, dirt and grease residue will be minimal. On the other hand, if you have just purchased a used classic, there may be years and years of accumulated road grime, dirt, grease, and debris hidden in the fenderwells, doorjambs, trunk wells, and engine compartment.

A complete day or two can be spent cleaning tires, wheels, and fenderwells. Another day can be spent under the hood and another on the interior. Only you can determine the specific time needed to clean and polish various parts of your car. In this section, concentrate on the preliminary wash and getting the vehicle clean down to a baseline. From there, the following projects will show you how to get past the baseline and into the finer points of detailing.

Work Area

For cars in relatively clean condition, the wash can take place in your driveway with little worry of surface stains and dirt buildup. For the others, consider using a self-serve car wash that is equipped to handle larger amounts of dirt and grime accumulations.

It is best to wash any car in the shade to minimize water spotting. Realizing this is not always possible, wash the shady side of the car first, saving the sunny side for last. Rinse frequently to prevent soap from drying on the body. If you must wash under a tree, be alert to tree sap and bird droppings. Wash them off immediately with cold water.

Consider parking your vehicle facing uphill. Most automobiles are designed to shed water from front to back. During rinsing, water will flow in the direction intended and easily run off moldings and trim and from door and sunroof drains.

The temperature of the vehicle's body is another consideration. You know what it feels like to jump into a pool of cool water on a hot sunny day. The paint on cars goes through a similar shock. If a car has been parked in the direct sunlight for over an hour, let it sit in the shade for a while until it cools. Feel the paint with your hand. It should feel no hotter than the ambient temperature. This will prevent "thermoshock" and the resultant creation of minute cracks that could threaten the longevity of paint. This is especially important for vehicles with new high-tech, clear-coated paint finishes.

Dismantling

For super washes undertaken once a year or for the first time, you should consider pulling the tires and wheels off the vehicle. This will give you excellent access to the fenderwells and the backs of tires and wheels. Do each wheel and fenderwell separately; you need not raise the entire car and do all four wheels at one time. *Do not attempt this unless you have jack stands or other means to adequately hold the vehicle with a wheel and tire off!* If all you have is a stock bumper or scissors jack, forget it. You must block the vehicle on a jack stand before you stick your head into the fenderwell.

Take everything out of the trunk, including loose carpet and mats. You will need to wash the spare tire and jacking equipment. The carpet can be vacuumed and shampooed later, if needed. The mats should be washed and dried.

Interior ashtrays should be removed, allowed to soak in warm sudsy water, and then cleaned and dried. Unusually

The place you choose to wash your vehicle should be convenient, allow for good water drainage, and have an adequate water source. Driveways are generally perfect, except for those times when vehicles are caked with mud, like 4x4s that have been out puddle jumping. For those, consider an initial rinse at a self-serve car wash.

Preliminary Washing

dirty dashboard knobs can be pulled off and soaked in the wash bucket. They will be cleaned later with a toothbrush. If the interior is in really bad shape, you should pull the seats out of the car. Necessary in extreme cases, removing the seats will give you more accessibility and working room to clean the interior thoroughly.

To prepare an automobile for its first Concours d'Elegance event, some owners will have their car dismantled to the frame. Each part will be cleaned, painted, and polished to perfection. This operation can take years, even for professional shops. You will have to decide how far to dismantle your car. The more you take off, the better the detailing job will be, if care is given to each piece and if the pieces are reassembled correctly.

Washing Equipment

Always follow the instructions on the label of any product you use. Generally, car wash soaps recommend that you place the soap in a bucket first and use a heavy water spray to bring up foam. Most labels also include a guide that shows how much soap to use per gallon of water with regard to the size of the vehicle you are about to wash. For larger cars, as an example, use two capfuls of liquid car wash soap per two gallons of water. Most car wash soaps are advertised as products that safely remove road grime without surface abrasion, eliminate hard water spotting, renew wax shine and protection, and inhibit corrosion and oxidation.

Some auto enthusiasts enjoy good results using mild liquid dish soaps to wash their cars, figuring that dish soaps that claim to be soft on hands must also be soft on paint. Others openly disagree, feeling that dish soaps and detergents are tough on wax and make quick work of removing otherwise nice wax finishes.

Fill your wash bucket about halfway with water, allowing soapsuds to fill the rest. One bucket of soap and water will not do the entire job. As you work through the preliminary wash, you will note that the wash water needs occasional freshening. Washing lower parts of a vehicle causes grit buildup in the bottom of the bucket. This grit will always find its way back to the wash mitt and the car's paint if you fail to rinse out the wash bucket occasionally.

After the wash mitt has had a moment to soak up water, dip it in the soapy foam. Use the foam for washing. This will reduce the amount of water and add cleaning agent to the mitt. It will also reduce the amount of water under the car while you wash underbody areas. Periodically, dip the mitt in the water, rinse thoroughly, and grab a new mittful of foam. If at any time you feel the mitt is impregnated with grit, use the garden hose to rinse completely. Empty the bucket and start fresh. Constantly keep in mind that you are rubbing on a precious surface, your car's paint. The slightest piece of grit in

Below Left: It may be a good idea to partially dismantle your vehicle in preparation for a detailed wash. Think about removing license plates, mounted spare tires, detachable racks, bras, and other easily accessible items to help in completing a thorough wash job.

Below Right: Along with a water source, garden hose and nozzle, you'll need a wash bucket, car wash soap, wash mitt, and drying towels or a chamois. Have a soft floppy paintbrush on hand, too. You'll use it to clean along window trim, body side moldings, and so on.

Detailing Techniques

the mitt will cause scratches on the paint, which will then require a rubout and rewax.

The preliminary wash may require the strength of a plastic-bristled brush around the fenderwells and on exposed underbody and tires. Paintbrushes work well on the grille, trim, and mirrors, while a toothbrush can be used on light lenses and rubber. Put the mitt and brushes in the bucket so they will be close at hand and not get kicked away. If the bucket is too small, put them out of the way on a grit-free surface.

A garden hose is perfect for rinsing. Before you begin, always make sure the hose will reach all parts of the vehicle. It is ideal to employ a hose with the metal end cut off, allowing water to flow freely and in large quantity. You don't need much pressure. Free-flowing water from an open-ended piece of hose will run off a waxed car body in sheets, leaving the surface almost dry. Nozzle spray will result in hundreds of small water droplets, making drying a bit more tedious.

If you are concerned about water conservation, and you should be, screw on a shut-off valve on the end of your regular garden hose. Then attach a short piece of hose, one with the metal male coupling end cut off. This way, you won't have to worry about the metal coupling scratching the paint. Free-flowing water will remove soap nicely and the shut-off valve will prevent water waste.

Washing a car requires no special skills. After a good rinse, dip the wash mitt into the bucket of car wash soap and begin washing the vehicle. Try to maintain complete coverage to eliminate missed spots. Start at the top and work down.

Instead of using an ordinary nozzle, consider using a small section of hose with the male coupling end cut off. Here, a shut-off valve was connected to the end of a hose and a smaller section of hose connected to it. Removal of the coupling eliminates a scratch hazard potential, and the soft flow of water will fall off the truck in sheets.

TECHNIQUE 1 BASIC VEHICLE WASH

Time: 2 to 6 hours

Tools: Water supply, bucket, wash mitt(s), scrub brush, paintbrushes, car wash soap, towels

Talent: ★★

Tab: $15–$25

Tip: Plan an entire afternoon for a comprehensive and thorough wash

Gain: Cars always seem to run better when they are clean

Complementary project: Clean the window glass on the inside

Above: Rinsing fenderwells will remove large accumulations of dirt, grime, and other residue. Concentrate on doing a good job of thoroughly rinsing each fenderwell. This includes the top, front, back, and the area of the body behind the wheels. You might be amazed at how much stuff is loosened up and floated away by a good rinsing.

Right: Whenever touching the paint finish on your car or truck, try to maintain a straight back-and-forth pattern, as is being done here with a wash mitt and car wash soap. Maneuvering in straight patterns helps to reduce swirls and spider webbing on paint surfaces.

Make sure your washing tools have been rinsed with clear water and are free of grit. Mix up a fresh solution of car wash soap and water and have both your newer and older wash mitts at the ready. Consider removing the license plates and anything else you want off the car for this project. Start by rinsing the vehicle first. Some car cleaning "experts" think high-pressure water is great for getting the big stuff off the body. Most others strongly disagree. They feel if the pressure is great enough to blow dirt off the vehicle, it is apt to push dirt into the paint. Water under high pressure can peel off paint and emblems, accentuate tiny paint chips, and possibly cause tiny scratches.

Start rinsing at the roof. Work your way down each side of the vehicle, making sure all of the big stuff is floated away. Then, rinse the fenderwells, tires, and wheels one by one. As you move from one wheel to the next, rinse off the underbody along the way. This includes both sides, the front and rear.

Most car enthusiasts like to initially rinse with a good flow of water from a nozzle. The pressure helps to dislodge grit from underbody areas and other loose debris from trim and molding. They wash the body with a soft cotton wash mitt and lots of soft, sudsy soap foam and then rinse with a soft flow of water from an open-ended section of hose. They treat the paint on their cars as if it were their own skin; realizing that paint is soft and any harsh treatment can result in blemishes.

Use the wash mitt in a straight, back-and-forth motion. Some car care product labels recommend application in a circular pattern. In some cases, a circular motion may cause spider webbing. You should try to wash, dry, polish, and wax your car in this same straight back-and-forth pattern every time. Massaging the body in this manner helps to prevent swirls. In addition, airflow goes over cars in the same direction. It is conceivable that small dust and dirt particles in the turbulence flowing over a car are less apt to get caught in tiny paint swirls if they are in line with the flow.

Start washing at the top of the car, washing one side of the roof first. Use a soft paintbrush to reach in the grooves along windshield and rear window trim. In the shade, on a cool day, you can also wash one half of the windshield and rear window at the same time. Rinse with a soft flow of clean water. Wash one half of the hood, using the paintbrush along the gap between the hood and fender, as well as around the radio antenna, hood scoops and windshield wiper blades. Rinse with clear water. Then wash the

Detailing Techniques

side windows and use the paintbrush along window trim, doorpost, and mirror. Move on to do one-half of the trunk in the same fashion. Be sure to remove all of the soapsuds each time you rinse.

Continue washing all parts of the car in the same manner. Use a paintbrush on every nook and cranny; it should work very well. Dip it in the wash bucket frequently to rinse clean and gather foam. If wax buildup is a problem on emblems or light lenses, use a toothbrush. Before washing each new area, freshen the mitt by briskly agitating it in the bucket. If at any time you feel the mitt is too dirty to continue, rinse it with the hose; rinse the bucket, too, because chances are good that it is also contaminated with grit.

A common pattern for washing cars involves starting at the top, then going around the sides to end up at the grille. If the lower sections of the vehicle are quite dirty, consider washing just the upper half on the first go around. Then, with a fresh bucket of soapy water, wash the dirtier lower sections separately.

The thought to keep in mind is completeness. Every part of the body must be cleaned. To do this, you need a method that guarantees you touch every square inch of the body. Use any system you wish, as long as it works. The more you use a preferred pattern, the less likely you are to miss spots, and this will cut down the time you have to spend going over the car looking for them.

Very dirty vehicles should be washed twice. The first wash will remove heavy stuff, and the second will include a more detailed cleaning.

As you wash the vehicle, take note of special items, such as pop-up headlights, chrome exhaust tips, and wiper blades. More than likely, the headlights on your Corvette or Porsche will be in the down position while it is parked. Wash the topside as you come to it. Then, before drying, hop in the car and pop up the lights. Use the paintbrush around trim rings and housings; rinse thoroughly.

Extended chrome exhaust tips are cleaned with the wash mitt. If necessary, use an SOS pad to remove carbon buildup. Later, you can polish with Happich Simichrome, Mothers, or Meguiar's chrome polish. The same holds true for bumpers located directly above exhaust tips. The bottom edge of these bumpers takes a beating from exhaust. The heat and products of combustion cause chrome to rust and deteriorate. It is important to clean and wax the bottom side of these susceptible bumpers on a regular basis.

Windshield wipers are probably the most neglected part of any car until winter, when you really need them. While washing the windshield, take a moment to wash the entire mechanism with the mitt. Use the paintbrush to reach inside the grooves. A toothbrush works well to remove bug residue. At the same time, look at the windshield washer nozzles. Frequently, they are clogged with polish or wax. Use a toothbrush first. If the tiny holes are plugged, you can use a needle to clean them out, although it is best to disconnect the hose from the windshield washer nozzle and blow out the nozzle with an air hose.

Every car has its own special features. It is up to you to notice and take care of them. Look at all the emblems and light lenses. These are very common spots for polish and wax buildup. Use the toothbrush to remove that residue. Check the edges of body side moldings. They catch a lot of debris, such as pine needles, small leaves, and lots of dirt. The paintbrush works well to remove that stuff. Also look at the edge of clear plastic mudguards. They tend to attract wax buildup that is removed with the paintbrush.

Since the jacking equipment should be out of the trunk, clean it after the car is dry. If your car is equipped with a tool kit, open it up. You may be surprised at the condition of the tools. Your pliers may be frozen, screwdriver rusted, and who knows what else. Use a toothbrush, 0000 steel wool, and WD-40 to get your tools squared away.

Clean the spare tire and ensure it is filled to its correct air pressure, too. Scrub the rubber if needed and wash the inside of the wheel. It is a good idea to wax both sides of the wheel and dress both sides of the tire, wiping off the excess. While you

Top: Carry a soft floppy paintbrush in your bucket of car wash soap. As you bring up a good solution of soapsuds along window trim and in tight spaces with the wash mitt, follow up with agitation of the paintbrush to loosen up and dislodge dirt and grime.

Above: As you work your way down the vehicle from the top, be sure to check the condition of the car wash soap and water in the bucket. Is it collecting grit or has it become dirty? If so, dump it out and mix up a fresh solution. As you wash the lower areas of the vehicle, check to ensure the wash mitt is not collecting a bunch of grit. If it is, rinse it out with clear water before dipping it into the bucket again.

Preliminary Washing

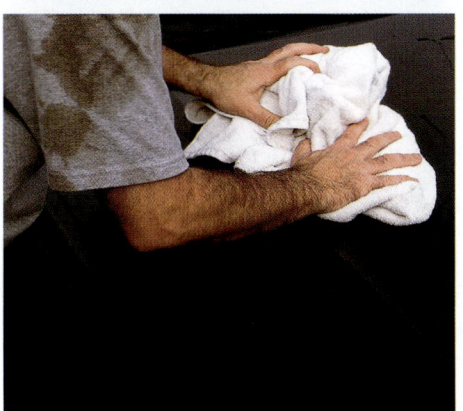

Top: Windshield wipers are frequently overlooked during most general car washes. After an initial cleaning with a wash mitt, use the paintbrush to reach into tight grooves and slots to clean off bug residue, dirt, and grime. Rinse with plenty of water.

Above: A debate between detailers has been ongoing for years regarding the use of soft cotton towels or a chamois for drying vehicles after a wash. Good results have been enjoyed using both, as long as both are clean to begin with. To reduce the amount of water normally left standing after a car wash, try using a squeegee first to remove the bulk of standing water. Follow with a soft towel or chamois.

are at it, wax the jacking equipment and give the tools a wipe with WD-40. These measures will prolong the life of the equipment while it is stored out of sight and out of mind in the dark of the trunk.

Automatic Car Washes

Most car enthusiasts share the same opinion of automatic car washes—they will hurt the car's paint and do not do as good a job of washing as can be done by hand. The sole beneficiary of automatic car washes may be a regular driver—with an owner who refuses to detail and is on his or her way into a dealership ready to trade the vehicle in on a new one.

Automatic car washes cannot remove wax buildup in emblems, decals, and trim. Even the softest brushes scratch paint. Many car washes promise and provide a good job but cannot possibly guarantee the same quality as a thorough wash by hand. Although a car wash may advertise scratch-free brushes, you must be concerned about grit removed from other vehicles that remains caught inside the nonscratch brushes.

Drying

Getting all of the water off a vehicle is almost as important as washing, especially if you live in an area that has unusually hard water. Water spots are unsightly and will require polishing and waxing for removal.

Some car folks prefer to drive their cars for a mile or two after washing. They believe that the movement of the car and the wind passing over it help to remove puddles of water from all the nooks and crannies. Others don't like the idea because they believe road grime will easily adhere to the wet sides of the vehicle after being thrown up by the tires.

Ideally, you could dry a car with a soft towel and then drive it on a superclean road to remove hidden water. This would help to prevent water spotting on the body and remove puddles that could encourage rust.

To illustrate why water removal is so important, consider this detailer's dilemma. A classic car owner seldom drives a classic 1960s vintage muscle car. The car is washed religiously every Saturday; a labor of joy and love, no doubt. After towel drying, the car is parked back in the garage, where it sits until the next Saturday. After a couple of years, the owner decides to take it out for a spin on a dry, sunny afternoon. To the owner's surprise, the brakes on this classic are nonexistent! Repeated washing, with virtually no driving exercise, allows water to sit idly on the brakes. After a prolonged period of nonexercise, the brakes will become so rusty and corroded they will fail to perform.

Drying a car is simple. Use two soft towels folded in quarters. The first one is used for primary drying and the second for picking up remaining streaks of moisture. As towels become wet, refold to drier sides. Use the same wiping pattern as you did for washing, straight back-and-forth motions. Start at the top and work down, then go from front to rear. Use folds in the towels to absorb water in tight spots around headlights, bumpers, and the like. Don't forget to dry the wheels and tires, too.

Car people seldom use a chamois. They feel that they streak and hold grit and grit particles. They prefer large, fluffy soft cotton towels. After cleaning, they put towels in the washing machine and clothes dryer to have them ready for the next time they are ready to clean their car.

To help speed the process and enable towels to stay dryer longer, consider using a squeegee. From the small assortment at an auto parts store, select one made specifically for car bodies. These types will include features that reduce scratch hazards, such as rounded edges, no metal frame brackets, and so on. Start at the roof and squeegee your way down the vehicle in much the same pattern you employed while washing. This will help to remove most of the standing water quickly and allow the towels to do a much better job of drying.

Detailing Techniques

TECHNIQUE 2 DOORJAMBS, HOOD, AND TRUNK EDGES

Time: 1 hour

Tools: Towels, cleaning cloths, all-purpose cleaner, paintbrush

Talent: ★

Tab: $5–$10

Tip: Turn down the water spigot when rinsing with a garden hose to reduce water splash

Gain: Clothes and things are not soiled when brushed across these edges

Complementary project: Lightly lubricate hinges after a thorough cleaning

Doorjambs

Use a damp towel to wipe down dusty doorjambs. If they have been maintained on a frequent basis, a few swipes with a damp towel should be all that is needed to remove dust and the small amount of dirt that generally accumulates along the bottom of the door opening. Go over both the door perimeter and the door's open edges.

Use an old wash mitt and a paintbrush to clean badly soiled doorjambs. There is no need to employ a lot of water and soap. A damp wash mitt will remove light dirt buildup, and soapy foam on a paintbrush will remove most other stubborn grime deposits.

A paintbrush will clean most of the dust and dirt from open vents commonly located on the latch side of doorframes. Don't spray much water on them. Unnecessary water inside these vents will filter down inside the door to become a source for rust deposits.

To rinse soapsuds from doorjambs, close the door and flow a small stream of water into the gap between the door and the body. You will be surprised how well this technique works. Afterward, open the door and remove lingering suds with a wash mitt or towel.

If you are washing on a warm sunny day, don't forget to rinse the parts of the vehicle body that get soapsuds on them. This will help to prevent the formation of water spots. Make sure you dry them, too.

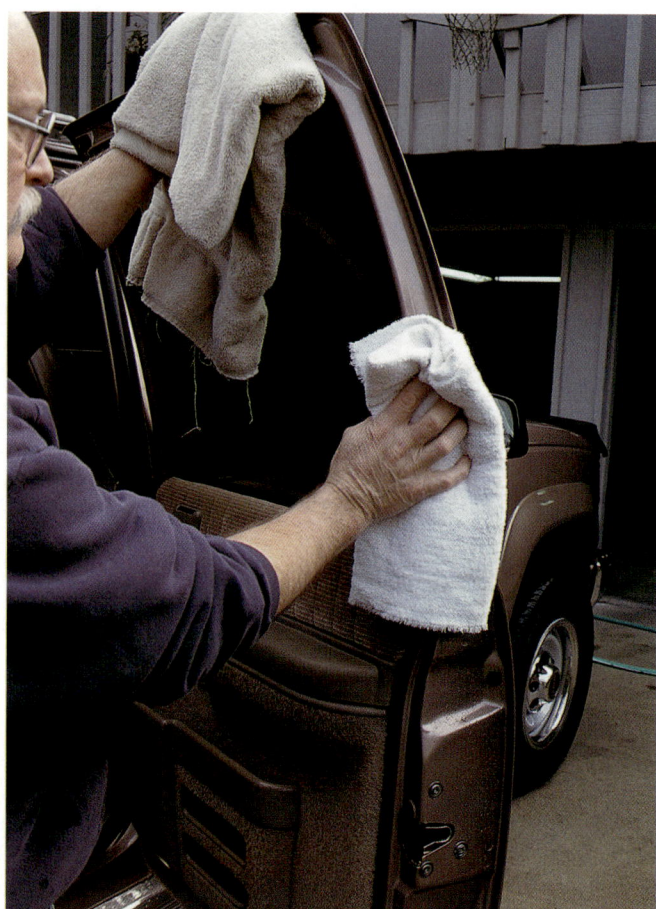

On a regularly washed vehicle, the drying towel seldom gets dirty. Use it in its damp state to wipe down door edges and the door opening perimeter. It should quickly remove dust and light films of grime.

Hood Edges

Pop open the hood and take a look at its perimeter edges. This part of the vehicle is usually marred with a buildup of dirt and crud. It is generally easy to clean lightly soiled edges with just a damp wash mitt, towel, or cleaning cloth.

For more stubborn hood edges, spray a little all-purpose cleaner on a cloth and wipe away the dirt. Consider using a wash mitt to protect your hand. Finish the job by going over the hood perimeter with a clean towel dampened with clear water.

After that's done, take a look at the engine compartment perimeter, which surrounds the hood edges. Use much the same technique to clean it. Employ a floppy paintbrush in tight spaces to agitate all-purpose cleaner and loosen up dirt and crud. Follow up with a damp towel to wipe away dirt and soapy residue. Be sure to wipe off and dry any parts of the fenders that were also moistened with soap or water.

Trunk Edges

Open the trunk and inspect the lid edges and opening perimeter. You will likely find a buildup of dirt along the front edge of the opening closest to the rear window. This is a common place for leaves, pine needles, and other loose debris to accumulate. Pull out the big stuff with your hands, being careful to avoid cuts and scrapes from sharp edges.

A damp wash mitt works great for getting into trunk opening perimeter grooves. After most of the dirt is removed, lightly spray all-purpose cleaner on a cloth and work on the tougher spots. Continue until the entire perimeter is clean. Wipe away cleaner residue with a water-damped towel.

Trunk lid edges are cleaned just like those for the hood. Inspect the rubber molding that goes around the lid edge to ensure it is in good shape. Cuts and tears in it can prove to be a source for water leaks into the trunk. Replace as necessary.

After wiping off the door edges, pop open the hood and wipe down its edges along with the engine compartment's outer perimeter. This will not only get rid of any standing water, but it will also help to wipe away dust and dirt that have accumulated along the surface.

Right: Wipe off the edges around the trunk lid and the trunk opening perimeter. If the towel becomes soiled, get a clean one. If the area around the trunk perimeter is really dirty, spray a towel or cleaning cloth with Simple Green and use it for cleaning.

Far Right: During the regular car wash, you should have opened the fuel filler door to wash that space. A paintbrush does a good job of removing dirt buildup. Don't forget to dry off that space with a towel, checking to make sure it is clean.

Detailing Techniques

TECHNIQUE 3 REMOVING TREE SAP, BUG SPLATTER, AND THE LIKE

Time: 1/4 to 1 hour

Tools: Paper towels, soft cotton towel, bug and tar remover, polish, and wax

Talent: ★

Tab: $1–$15

Tip: The sooner you attempt to clean off this stuff, the easier it will be to remove

Gain: Less likely chance paint underneath will be permanently marred

Complementary project: Clean off road tar and other junk from lower body areas

Tree sap, bird droppings, and insect remains are common problems. The best advice is to remove them as soon as possible. If after a wonderful dinner with a special friend you return to your ride to find a gargantuan sea gull has dropped last week's entire menu on the hood, clean it up right away. Go back into the restaurant, get some wet paper towels, and clean it off.

Tree Sap

Tree sap can sometimes be removed with just cold water. Gently rub the spot with a soft towel moistened with clear cold water. It might easily dissolve after a while. Apply a glaze polish after the water and towel treatment to ensure all of the residue is removed. Use a stronger polish if necessary. Follow with a light coat of wax.

For tree sap that has been on the surface for an extended period of time, you may have to employ a stronger cleaning method. Many of the bug and tar removers on the shelves at auto parts stores will do a good job of loosening up tree sap. Be sure to follow directions and let the solution sit on the spot for a long enough time to allow the ingredients to start dissolving. Plan to go over the spot a number of times, as seldom does tree sap come off on the first try, especially when it has hardened all the way through.

Should none of the above methods seem to work, talk with the clerk at your local autobody paint and supply store. He or she may carry a more potent product that can do the job.

In extreme cases, you might consider carefully using a razor blade to slowly and patiently scrape off a tiny layer of sap at a time until you get close to the paint surface. Mask off the tree sap spot with tape to help protect against an accidental slip with the razor blade. Once you get close to the paint surface, employ a bug and tar remover, adhesive remover, or other such product to get rid of the remnants.

After an aggressive attack on hardened tree sap, wipe off the area with a damp towel and then polish the spot and surrounding area with a mild glaze. Follow that with a light coat of wax.

Bug Splatter

Some bug splatters are easy to remove from painted surfaces and brightwork with just soap and water. Others require more time and elbow grease. Tough blemishes on chrome are easy to remove with number 0000 steel wool and a little

1. A part of this painted bumper (above and left of the license plate) was blemished with bird droppings. Allowed to settle on a painted surface for any length of time, bird droppings can permanently damage paint. Make every attempt to clean off bird and insect droppings as soon as possible.

2. The area of the bird dropping is washed with a mitt dipped into a bucket of car wash soap and water. The dropping material was easily washed off with a few applications. Rely on more than one application with soap and water before resorting to more abrasive means to remove bird droppings, tree sap, and the like.

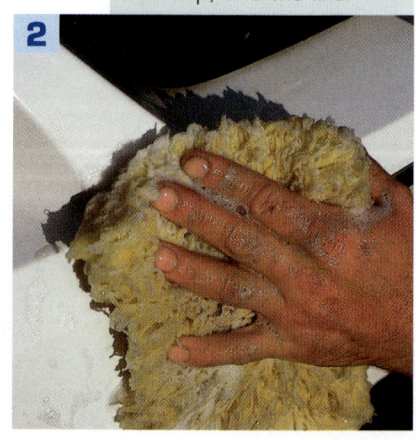

Preliminary Washing

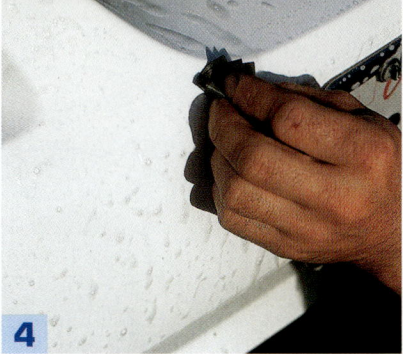

3. Although a clay bar could have been tried, this detailer is employing a very fine 1,000- to 2,000-grit wet/dry sandpaper to work out the blemish caused by the bird dropping. Notice that the area being wet sanded is very wet, an important step to remember when wet sanding.

4. Notice how the blemish has disappeared with use of the very fine wet/dry sandpaper. Use very light pressure on the wet/dry sandpaper when wet sanding. Opt for a number of very light passes as opposed to a single heavy one.

5. With the bird-dropping blemish removed, a sealer/glaze polish is used to remove the light sanding scratches left behind from wet sanding. Polish can be applied with a clean cloth or a damp applicator sponge. Wipe off dried polish with a clean, dry, soft cotton cloth.

6. The results of a little elbow grease and patience have paid off. The blemished spot of paint where the bird dropping was located is now clean, polished, and looking great. That spot should now be protected with a light coat of wax.

all-purpose cleaner. Just dampen the steel wool with cleaner and rub away.

On paint and nonchromed parts, however, you must take your time and employ less abrasive methods to remove dried-out bug splatter. Try soap and water first, knowing that it will require repeated applications; as many as 10 to 15 different efforts—rub the spot with a sudsy wash mitt, rinse the mitt, do it again, and so on. If this just doesn't work, graduate to the bug and tar remover of choice.

Follow the instructions on the product you employ. Apply the remover to a cool vehicle surface. Gently work it into the splatter with a soft cotton cloth. Allow the solution to sit for about 30 seconds, and then wipe it away with a clean cotton cloth. Follow up with a soft cloth dampened with clear water. Be advised that you might have to apply the product more than once to ensure all bug splatter residue is completely removed.

Insect droppings are removed with the same care and patience. By all means, attempt to remove this material as soon as it is noticed. The longer these acidic residues stay on your car, the deeper the paint damage they can cause. Try the mildest soap and water method first and resort to more potent methods as necessary.

Other Blemishes

Road tar is just one of many paint hazards your automobile is faced with every time it is taken out on the road. Not too terribly noticeable on darker colors, it can stand out like a sore thumb on white and lighter colored vehicle bodies. Soap and water are generally no match for this stuff, so just prepare yourself for the fact that you are going to have to apply some effort in getting rid of it.

Bug and tar removers work very well to quickly dissolve road tar. Many light polishes also do a good job of removing these black spots. If you have chosen to use a combination cleaner wax product on your vehicle on a regular basis, you might be surprised as to how well it also works to remove spots of road tar.

Of greatest importance when using any of these road tar–removing methods is to remember to unfold the cleaning cloth to a clean side frequently. Don't just saturate one side of a cloth with remover or cleaner wax and expect it to be able to handle all of the road tar from an entire side of the automobile. You will just end up smearing absorbed road tar all over the next section you try to clean. Work on a small section at a time and then unfold the cloth to a clean side. Have another clean soft cotton cloth ready for a final wipe down after the tar is removed from a section.

If you are hesitant about using a bug and tar remover product on your car's paint, apply it first to an inconspicuous spot down under a bumper or on a bottom door edge. Although products made by companies like Meguiar's and Mothers are designed for use on automotive paints, you can reassure yourself of their performance without worry by testing them on a hidden spot.

The bottom line to keep in mind whenever attempting to remove tough contamination from paint is to start gently and progress to more aggressive methods slowly. Uninformed car owners have used plastic scrub pads to remove bug splatter and the like from their vehicle's paint. In the garage under a light, the clean spots initially looked fine. Out in the sun, however, the scratches left behind were most obviously apparent.

TECHNIQUE 4 GRILLE, BRIGHTWORK, AND RUNNING BOARDS

Time: 1/2 to 1 hour

Tools: All-purpose cleaner, cleaning cloths, toothbrush, paintbrush, number 0000 steel wool

Talent: ★

Tab: $1–$5

Tip: Have patience to prevent smears or scratches on nearby paint

Gain: Helps to accent an entire vehicle

Complementary project: Clean and dress surrounding vinyl and/or rubber trim

Grilles

Clean off surface dirt and road film from wide grille patterns and emblems with a cleaning cloth and an all-purpose cleaner. Use a paintbrush to agitate cleaner around headlights and for reaching into recessed areas. Duct tape wound around the metal part of the paintbrush helps prevent paint chips and scratches in these tight spaces. Stubborn bug residue can commonly be removed from grille work with water and a toothbrush.

Grilles present detailers with lots of tight areas that are generally filled with dust and dirt. A wash mitt or cleaning cloth will only reach partly into them, leaving unsightly spots. A soft floppy paintbrush works great in and around these confined areas. Take your time cleaning and rinsing to assure yourself of a thorough cleaning job. Use a toothbrush or cotton swab, if necessary, to remove light dirt remnants tucked away in recessed corners.

If the grille on your automobile presents a host of small rectangles and other hard to reach spaces, consider wearing a pair of cotton gloves and using your fingers as active cleaning tools. You can dip your glove-covered fingers in soap and water or lightly spray them with cleaner to simply massage dirt away. Gloves will also help to protect fingers from cuts and scrapes.

Brightwork

Brightwork is a term used to describe just about any shiny metal piece of automotive trim. Although commonly regarded as those parts that make up front

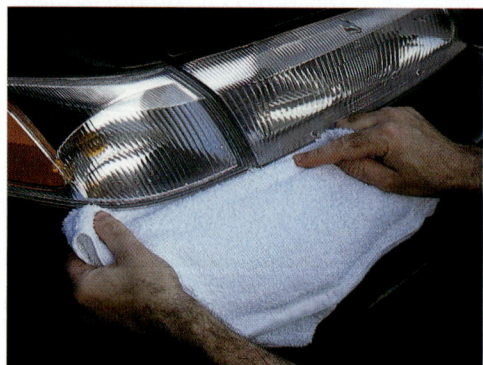

A towel has been folded and slipped into the tight space between the light and bumper. It will wick up standing water to prevent the formation of water spots later on. Brightwork needs to be clean and dry before it is polished.

The chrome and brightwork on the grille of this vintage automobile is stunning. Meticulous care was given to each piece as it was patiently polished. Washing it after a drive will require careful attention to all the nooks and crannies, where bug residue and road grime accumulate. This is where a soft, floppy paintbrush will come in handy.

Preliminary Washing

Above Left: Brightwork includes just about every shiny piece on a vehicle. A soft, damp cloth is used to clean the chrome strip around the convertible top boot. It is followed with a few swipes with a dry cloth.

Above Right: Running boards must be washed with the rest of the vehicle. Rubber pads attached to running boards are cleaned with car wash soap and a plastic-bristled scrub brush. A toothbrush may be needed for cleaning between tight grooves and along edges between the rubber and chrome.

grilles, brightwork can include shiny light housings, trim pieces, hood ornaments, chrome bumpers, and so on.

During a normal car wash, most brightwork is touched by a wash mitt and cleaned to a degree. However, upon closer inspection you will likely recognize pockets of dirt and other buildup missed by the bulky mitt. A thorough cleaning generally requires the use of a soft paintbrush and/or toothbrush.

As you go over each piece of brightwork with a soft cloth and cleaner, look closely for buildups of polish, wax, bug residue, and other stuff. Spray some cleaner on the bristles of a soft toothbrush and then gently whisk away dirt residue. This is especially important on pieces that feature grooves or slots. In some cases, such grooves become so caked with dirt that they may appear to be painted. As the toothbrush bristles begin to dislodge dirt, you may be amazed to see how dirty the cleaning residue becomes and how bright the groove appears. Wipe away cleaning residue and any remaining smudges with a clean cloth or towel.

Badges, emblems, and other ornaments are also cleaned in much the same manner. Take your time and keep a clean towel handy to wipe off any cleaning solution that runs onto painted surfaces. Periodically, use a soft cloth dampened with clean water to wipe off cleaning residue from brightwork and surrounding areas. Don't be surprised if you find yourself spending a lot of time cleaning every piece of trim and brightwork all around your car. On vehicles with such pieces that have been neglected for some time, the cleaning results are commonly so graphic that detailers have a tough time stopping.

Running Boards

A lot of SUVs and 4x4s feature running boards for both usefulness and appearance. Not much different than other parts of an automobile, these items are most often cleaned with soap and water and a wash mitt. For more thorough cleaning, however, you'll have to be prepared to bend down and use a little elbow grease.

Plain pipe-like painted running boards are cleaned with a soft towel and an all-purpose cleaner. Use a soft paintbrush, toothbrush, or scrub brush to dislodge dirt buildup from around carriage bolts and other fasteners. A plastic scrub brush works best on nonskid strips to get dirt out from around the tiny crevices.

Chrome and aluminum running boards are also cleaned with a towel and cleaner, while fine steel wool may be used with a liquid cleaner on real chrome, if necessary. Use the paintbrush and toothbrush as necessary and consider a plastic scrub brush for those running boards with grooves and vinyl inserts. You may have to spray them with a cleaner like Simple Green and briskly scrub with a brush to break loose dirt and crud from the corners. Clean all around the running boards, including the front and rear edges.

TECHNIQUE 5 TIRES AND WHEELS

Time: 1/2 to 1 hour

Tools: Bucket of car wash soap and water, cleaner, wash mitt, brushes, scouring pads

Talent: ★

Tab: $5–$15

Tip: Roll vehicle back to expose lowest part of tire for cleaning

Gain: Crisper-looking tires and wheels

Complementary project: Clean the fenderwells

Tires

At the auto parts store, you will find an array of tire cleaning products, all claiming to be the best and easiest to apply. These products are designed as step-savers, some claiming to clean tires with minimal rubbing and scrubbing. Label information may profess that a unique formula will instantly penetrate and dissolve grease, oil, and road grime, while not using harsh petrochemicals, bleaches, or abrasives. You'll have to be the judge as to how well they work for you. Whitewall cleaners, like Westley's Bleche-Wite, can also be used for blackwalls. Watch out for cleaners containing bleach, as they may stain unprotected aluminum and alloy wheels.

A lot of car people have found the safest way to clean tires is with regular car wash soap, a cleaner, water, and a plastic-bristled brush. Simple Green is a product that will not harm metal, and it works very well to help get tires their cleanest. Using dish soap on tires often yields good results. Extremely dirty and stained rubber has come clean with powdered cleansers and a brush. Once again, you must remember that bleach in the cleanser can damage your aluminum or alloy wheels.

Rinse the fenderwell thoroughly. Dislodge accumulations of mud, leaves, road salt, and pine needles. Remember, these pockets of debris retain moisture, and they are common spots for rust to grab hold. Afterward, rinse the wheel and tire.

Start out with a bucket of car wash soap and water, a spray bottle of Simple Green, a scrub brush, and a toothbrush for complete tire cleaning. Rinse the tire, spray it down with Simple Green, dip the brush in the soap solution, and start scrubbing. Rinse and do it again, concentrating Simple Green spray on stubborn dirt spots.

Use a toothbrush along the rim and in tight cracks next to lettering and tread. If need be, use a liquid cleanser like Soft Scrub for even more cleaning punch.

Whitewalls can be scrubbed with a whitewall brush or an SOS pad, as both work well. The soapy solution in your wash bucket will add to the cleaning power of the Simple Green spray. Rinse with clear water. Be prepared to go over

The tire is initially washed with car wash soap and a scrub brush to bring up the lather. Here, the detailer is cleaning the whitewall with a washcloth. Depending on how dirty the whitewall is that you are cleaning, you may need to employ the strength of a whitewall brush. To help get the whitewall and the rest of the tire as clean as possible, a powdered cleanser can be sprinkled over the surface. A scouring pad can be used to further brighten the whitewall.

Preliminary Washing

Above Left: Wheels are easiest to clean with cleaner and a paintbrush. Simple Green works well, as does Meguiar's Extra and other all-purpose cleaners. After a few swipes with a wash mitt, use a paintbrush to reach into slots, grooves, fins, and other hard-to-access areas.

Above Right: Be sure to thoroughly rinse tires and wheels after each cleaning application. This will prevent soapsuds from drying on the surfaces and also float away dirt and grime to give you a better view of your progress.

white areas more than once for optimum cleaning results.

If the tire is unusually dirty and stained, you can use any number of tire or whitewall cleaners. Most of them work quite well, although a few users feel some whitewall cleaners tend to fade the black after repeated use. SOS pads offer good results, as do powdered cleansers when used along with a scrub brush. Keep in mind that the rougher the cleaning application, the greater chance you take of damaging a tire or wheel. Powdered cleansers contain a bleaching agent. Will that hurt the wheel on which the tire is mounted? Overall, it may be much better to clean a tire three or four times using a mild cleaner than once or twice with a harsh one.

Wheels

Few if any wheels are prone to damage when cleaned with car wash soap and water. Use an old wash mitt, rinsed and clean, to wash off dirt and road film from wheels. Use paintbrushes to reach those areas around lug nuts and valve stems and in the crevices and slots on specialty wheels. Toothbrushes work well on wire wheels and on those parts of the wheel that attract dirt buildup, like valve stems and balancing weights. Wheels designed with fins and slots are sometimes easiest to clean using your fingers. These handy little cleaning tools won't scratch or wear out. Consider wearing a pair of soft garden-variety cotton gloves to protect your fingers against cuts and scrapes. You can get these gloves wet with sudsy water from the wash bucket and use them as you would a regular wash mitt.

Before you leave the wheel, take a look at the fender lip, that part of the painted fender that curls around to meet the fenderwell. Wipe it with an older wash mitt. If need be, use a plastic-bristled scrub brush to clean off stubborn deposits of dirt and grime.

Wheel covers and hubcaps are also cleaned with a wash mitt, soap, and water. Use a soft brush to whisk away dirt buildup from grooves and crevices.

Always be sure to thoroughly rinse the fender lip, fenderwell, tire, and wheel before moving on. The clean rinse water will remove leftover dirt and soap residue. Generally speaking, you can never rinse too much.

After tires and wheels have been washed and dried, consider a light coat of dressing. Armor All, Meguiar's Endurance, and Mothers Preserves all work well. Be sure to buff off excess dressing for best results. Tire dressing can sometimes leave behind a blue hue on whitewalls. Clean it up by lightly wiping with a dry scouring pad.

TECHNIQUE 6 VINYL TOPS AND CONVERTIBLES

Time: 1/2 to 1 hour

Tools: Bucket of car wash soap, wash mitt, cleaner, scrub brush, toothbrush

Talent: ★★

Tab: $5–$10

Tip: Don't aim water spray directly at the plastic windows or ragtop edges

Gain: A crisp look and added longevity to vinyl or convertible tops

Complementary project: Apply protectant to vinyl top; lubricate ragtop hinges

Washing vinyl tops is no different than washing anything else, unless the top is terribly dirty, with buildup trapped in the grain. If the wash mitt doesn't clean to your satisfaction, use a plastic-bristled brush. Note that after such scrubbing, you must restore the vinyl with an all-purpose or vinyl top dressing. Scrubbing will remove most of the dressing already on the vinyl. Unless you apply a protectant, bare vinyl will be more susceptible to the sun's rays and will dry out, aging the material unnecessarily.

Convertible tops present no unusual problems except for leaks and the delicacy of the plastic windows. Ragtops are not watertight, especially around side windows. When washing, don't spray water directly at the tops of windows. Hold the hose at the roofline and let water flow down to prevent leaks inside the car. Use a minimal amount of hand pressure to wash and dry plastic windows. The material scratches easily. Meguiar's, Mothers, and other car care product manufacturers make products designed to polish these plastic windows and remove those slight scratches. However, don't rely on them to the extent that you are careless.

1. Vinyl and convertible tops are washed right along with the rest of the vehicle, using regular car wash soap and a wash mitt. Small stubborn spots of dirt and grime are removed with a toothbrush and a convertible top cleaner; all-purpose cleaners also may be used.

2. If spot cleaning fails to remove tough stains and dirt accumulations on vinyl and convertible tops, consider using a powdered cleanser with a scrub brush. Opting to try a powdered cleanser to clean a neglected vinyl or convertible top will require you to protect the painted finish with a sheet of plastic. This will reduce the chances of paint blemishes caused by the bleaching agents in cleanser.

Preliminary Washing

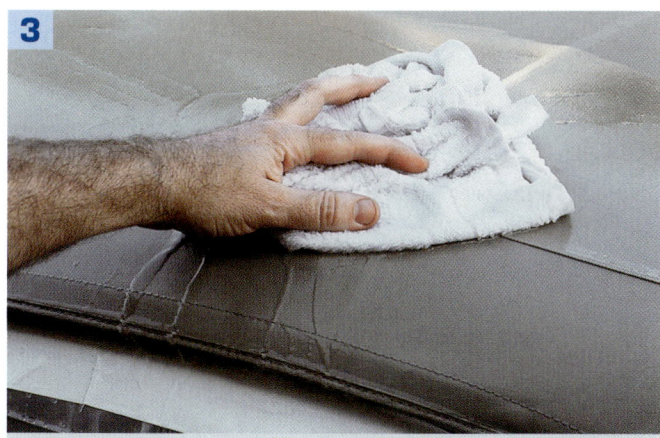

3. With the vinyl or convertible top wet from rinsing, sprinkle some cleanser over a small section of the surface and work it in with a wash mitt or cleaning cloth. Do only a small section (about 2 square feet) at a time and then rinse with plenty of water.

4. Should the wash mitt or cleaning towel fail to clean as desired, use a stout scrub brush to work in the powdered cleanser. Maneuver the brush in a circular pattern, reversing the direction to ensure complete coverage and maximum cleaning strength.

5. Be sure to scrub the edges of the top, too. Areas along seams and flaps are prone to dirt and grime buildup. Use a smaller brush as necessary to get into the tight spaces. Use caution while scrubbing next to the tops of windows and along the edge of the plastic sheet. Aggressive scouring could cause the sheet to tear or come loose, or cause the masking tape to loosen and fall away.

6. After scouring with a scrub brush and powdered cleanser, rinse the vinyl or convertible top with plenty of water. Use a cleaning cloth or towel to remove water and lingering soapsuds. If some spots on the vinyl or convertible top still present stains or accumulations of dirt after an initial scrubbing, clean them again with the powdered cleanser and scrub brush.

7. After a final rinse and thorough inspection prove the vinyl or convertible top is clean, use a clean towel to dry. Pay particular attention to seams and flaps to ensure all standing water is removed. Once the convertible top has been cleaned to perfection, pull off the plastic sheet, and wash the rest of the car. After everything has had a chance to dry, drop the top and clean the bonnet that covers it in the down position. Do not store a wet convertible top in the down position, as it will be most susceptible to mold and mildew.

TECHNIQUE 7 REMOVING STICKERS AND DECALS

Time: 1/4 to 1 hour

Tools: Adhesive remover, hair blow dryer, razor scraper

Talent: ★★↗

Tab: $5

Tip: Holding a hair blow dryer too close to a body surface for too long will damage paint

Gain: Turns a regular driver into a nice-looking automobile

Complementary project: Clean all window glass

Some body emblems are easily removed. If this is the case with your vehicle, take them off; it will make cleaning, polishing, and waxing much easier. If they are not easily accessible, or are otherwise difficult to remove, leave them on. There is no need to chance damaging the emblem or scratching the paint. License plates can come off and windshield wipers can be pulled out from the glass, out of the way.

Bumper stickers are designed for bumpers, hence their name. Under no circumstances should a bumper sticker be placed on a painted surface—especially a trunk lid. The longer they stay in place on a painted surface, the more paint may come off when one tries to remove them. Paint underneath will not match the rest of the car, because it will not have oxidized the same. Sticker removal could result in other paint blemishes that will require repainting for repair.

If you just purchased a car that is marred with a sticker on a painted surface, use caution while attempting to remove it. Products designed for decal and sticker removal are found at most auto parts and autobody paint and supply stores. These adhesive removers contain chemicals designed to loosen the glue employed to adhere the sticker to a surface. Take your time, go slowly, and peel just a little at a time. Impatience will cause errors that may result in paint

1. A hair blow dryer works well to loosen the adhesive used to keep stickers in place. Go slow and be aware that too much heat could damage paint. Heat an edge of the sticker and try to peel it away. As it loosens, pull more of the sticker away from the surface until it is free.

2. As the adhesive starts to loosen, use a fingernail to loosen an edge and start peeling the sticker off. As the sticker becomes tougher to remove, spray some more adhesive remover onto the exposed back of the sticker next to the painted surface.

Preliminary Washing

damage. Before you forcibly peel some of the material back using your fingernail, be sure the glue has loosened up.

Sometimes glue on the outer edges will give way while glue in the middle of the sticker is still firmly attached to the paint. Spray more adhesive remover onto the exposed side of the sticker. Let it stand for a few moments before continuing. Always read the instructions on the labels of any products you use. The engineers who designed the product likely spent many hours researching the effects of the chemicals, and their instructions on the label are the result.

A hair dryer can be used to loosen glue to remove stickers and decals. You must be very careful not to heat the paint to such a high degree that it becomes damaged or even starts to melt. As with chemical adhesive removers, take your time and work slowly to ensure a good job.

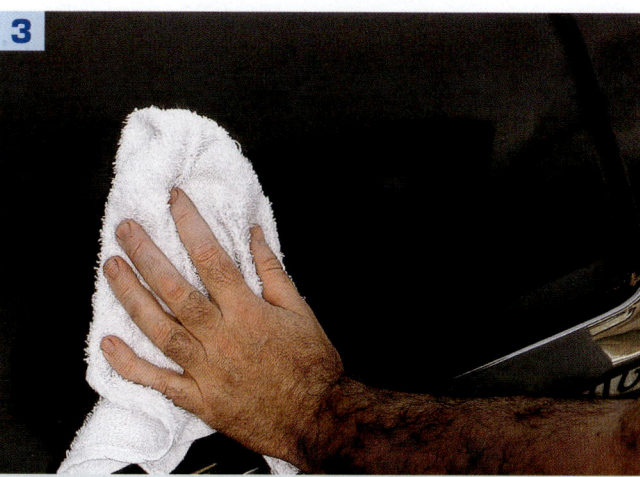

3. Wipe off lingering material with adhesive remover and a soft cloth. If sticker residue has dried on the surface, spray it again with adhesive remover and let it stand for a few minutes. Continue the process until the area is free of all sticker material and lingering adhesive.

4. In some cases, removal of a sticker from a painted surface will result in paint blemishes. Polishing the spot should improve the paint's appearance. Here, a detailer uses a polish designed for machine use along with a buffer outfitted with a foam pad. Smear polish around the area with the buffer off to eliminate excessive polish splatter.

TECHNIQUE 8 UNDERBODY

Time: 1/2 to 1 hour

Tools: Garden hose, car wash soap, wash bucket, old wash mitt, scrub brush, cleaner

Talent: ★↗

Tab: $5–$10

Tip: Don't forget to wash trailer hitches

Gain: Better-looking ride when viewed from other drivers sitting in cars

Complementary project: Paint fenderwells and visible underbody

Rear Underbody

True car enthusiasts wash their entire cars, including the visible underbody. Many like to start the wash job at the rear underbody. Because the vehicle will be facing uphill, water will run toward the rear of the car. First, rinse the entire underbody at the curb, let it drip dry, and then relocate to a dry spot on the driveway. This will give you a dry surface to lie on. Use a systematic approach. Wash the rear underbody first, because the ground will be dry and you won't get wet while crawling underneath. Then wash the side underbody and move toward the front. By the time you are done, the ground will be wet and you won't.

Fill the wash mitt with plenty of soapy foam. Use the foam to wash the visible parts of the axle, shocks, fuel tank, trunk bottom, and anything else around that area.

Use an old wash mitt for this chore. Use it for all the really grungy spots, saving your newer mitt for the body. This washing should loosen most of the dirt and road film that has accumulated since the last wash. If need be, wash the entire area twice before rinsing. Once you have rinsed with the garden hose, your work area will be wet and uninviting for you to crawl on.

Some car people place lawn sprinklers at the curb, turn on the water, and drive their cars over the spray to rinse the underbody. If you maneuver a car so that the sprinkler spray has access to all parts of the underbody, it should flush out some deposits of road salt, mud, and grime. You will have to drive over the sprinkler several times. One might surmise that this preliminary step could eliminate the need for as much elbow grease when serious hands-on cleaning begins. You be the judge.

Side Underbody

The same "dry" philosophy mentioned for the rear underbody holds true for washing side underbody areas, too. Since the rear area was done first, the ground under the sides should be dry. Get down on your knees, fill the mitt with soapy foam, and wash visible parts of the frame, exhaust, and body. Use the plastic-bristled brush as necessary.

Accumulations of road tar can be removed using any number of bug and

Thoroughly rinse fenderwells to remove accumulations of road grime, leaves, and crud. Use a brush to knock loose stubborn stuff and rinse it away with water. After a fenderwell has been rinsed, move along the side of the vehicle to rinse the visible side underbody.

Above Left: Work your way around the vehicle, rinsing all of the visible underbody. Be sure to get down low so you can see what you are doing.

Above Right: At a dry place away from where you did the initial rinsing, use a bucket of car wash soap and water and a scrub brush to clean the visible underbody. Try to load up the brush with soapsuds and little water to help keep the workspace as dry as possible while you are lying down on it.

tar remover products. Auto parts stores generally carry a good assortment. These products are specifically designed to remove road oil, tar, tree sap, dried bug residue, and the like. They are advertised as uniquely formulated to quickly and easily remove stubborn exterior stains, but you should still plan on such jobs taking longer than anticipated. As with any other product read label instructions before applying to your car's paint. When in doubt, try it on an inconspicuous area first.

Although the lower sides of your car may be hardly visible to you, take note of cars parked at stoplights and those on used car lots. Rust-colored frames extending from the bottoms of side underbodies are truly eyesores. Imagine how much better your car or truck will look with this area cleaned and painted black. It will add a crisp appearance to the entire side of the vehicle.

After a good cleaning using lots of soapy foam, rinse with clear water. Start rinsing at the front and aim the water spray toward the rear, leaving the area under the front end dry. This maneuver is designed to keep the detailer dry while cleaning under the front of the car.

If you have to go back and rewash a section of the underbody while the ground is wet, try kneeling on the hose. This will help to keep your knee out of the water and your pants dry.

Front Underbody

If the front of your car is equipped with a painted spoiler under the bumper, don't wash it with the same bucket of wash water used under the rear and side underbodies; there will be too much grit in it. Clean the spoiler with a fresh mix of soapy water before tackling other exposed front underbody areas.

Under the front, use the same technique to wash exposed steering members and the frame, radiator housing, and such. If greasy parts on the engine block or oil pan are plainly visible, use a mild solvent or degreaser. Gunk in the spray cans works well. Remember, if you are working on your driveway, don't get too carried away degreasing. The grease will fall on the concrete and make a mess. For large degreasing jobs, go to the self-serve car wash or other facility. You can put rags under the greasiest areas to contain small jobs; cardboard and newspapers also work but can be more difficult to clean up. It's simplest to wet a rag or paper towels with degreaser and wipe parts clean. Then wash with soap.

After rinsing, the underbody should be clean and your work area wet. If the job was done right the first time, you should not have to kneel down on the wet ground again. The rest of the wash can be done in a standing or stooping position.

Chapter 3
Interior Detailing

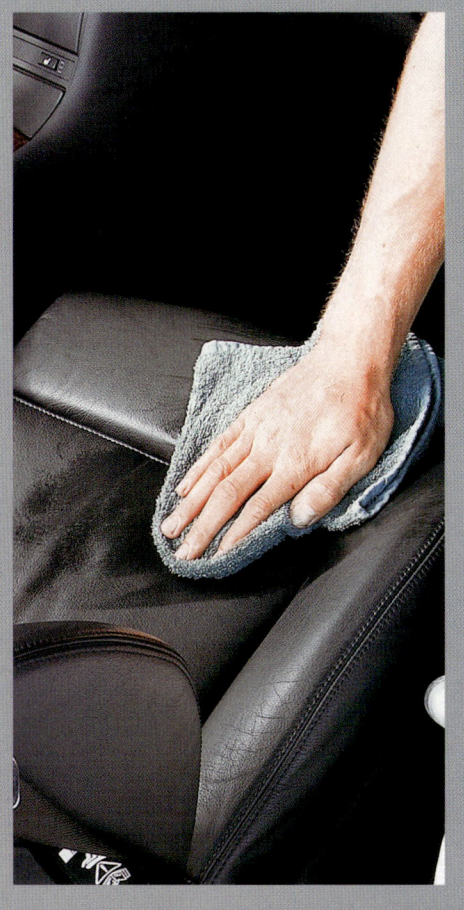

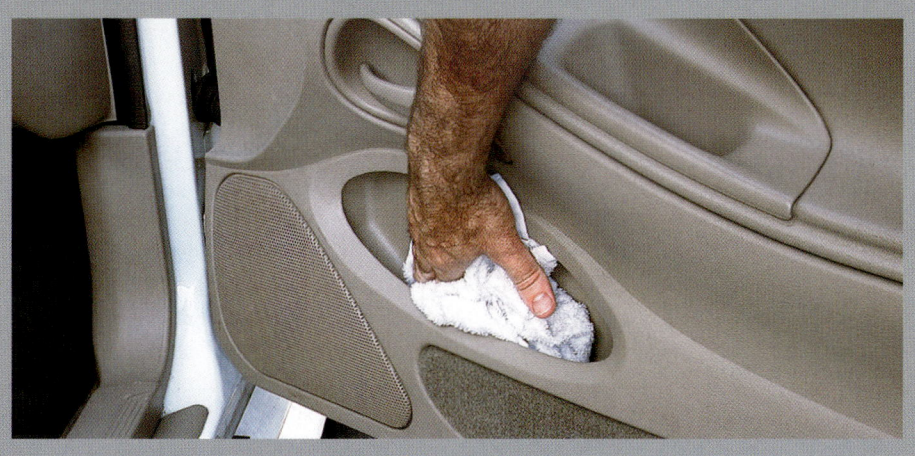

You are the one who needs to be happiest with—and most comfortable in—your car's interior. Here's how to make it perfect.

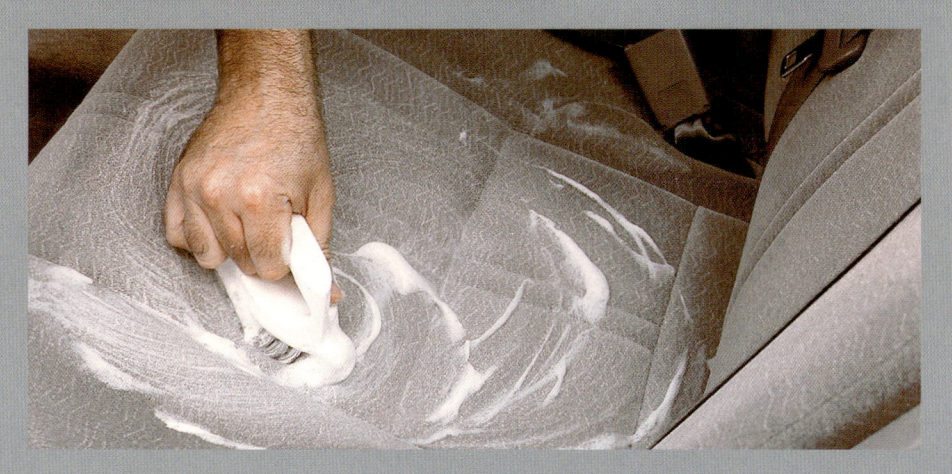

Detailing Techniques

Detailing a fine automobile is labor intensive, and in some cases a labor of love. Why else would anyone use a paintbrush and toothbrush while washing their car's exterior? The same basic logic holds true for interiors. Performing a quality job and pampering your vinyl, cloth, and leather upholstery will require time and patience.

Interior Dismantling

Cars need interior dismantling if they have sat in a body shop for extensive body and paint work or if their interiors have been neglected for a lengthy period of time. Basically, this means removing seats, easily detached knobs, consoles, and rubber pads on brake and clutch pedals. Clearing the interior makes work easier and gives you more working room. It also allows easier and more thorough access to the detached parts for cleaning.

If the inside smells like mildew, you might consider pulling the carpets. Chances are the carpet and padding are damp and the floorboard rusty. In these cases, dry the carpet and pad by hanging them out in the sunshine. If the pad dries without an odor, good. If not, consider replacing it. Carpeting should be shampooed and dried. If it is still in good shape, the mildew smell should dissipate, and you can reinstall the carpet.

A rusty floorboard should be cleaned, sanded, primed, and painted. The extent of rust damage will determine your course of action. If the metal is eaten almost completely through, you'll have to have a body shop repair the damage. If it is only surface rust, use wet-and-dry sandpaper to remove it and to expose a clean surface. Dry and then protect with an appropriate primer and paint, allowing each coat to dry before applying more. Check with a local body shop if you are unsure of the conditions and required repairs.

For interiors in better condition, your concern will likely be dust and grime buildup. With the seats, console, and knobs removed, you can vacuum every square inch of carpet, reach the tiny nooks and crannies on side panels, remove each speck of dust on the console, soak grime off knobs, and scrub seats entirely. You can also clean, repaint, and lubricate seat tracks. Do this while everything is wide open and easy to access.

This interior looks great. The carpet is clean and spot-free, the dashboard is clean and lightly dressed, the seats are tidy, the glass is clear, and everything looks crisp. It takes time and patience to make an automotive interior look this good, but it is well worth the effort.

TECHNIQUE 9 VACUUMING

Time: 1/4 to 1/2 hour

Tools: Vacuum cleaner, attachments, clean plastic-bristled brush, toothbrush

Talent: ★↙

Tab: $0

Tip: Use a crevice tool and soft brush attachments

Gain: Removes grit that can work to wear away carpet fibers

Complementary project: Shampoo carpet

A thought-out, systematic approach to interior detailing will save you time and result in a quality job. Vacuuming is a good place to begin. Removing the big stuff first provides a feeling of accomplishment and should boost your spirit to continue.

Use a crevice tool on the end of the vacuum cleaner hose. A plastic crevice tool is preferred because it will not pose a potential tearing hazard, as a metal one might. Through extended use, the metal tool often bends or tears at the end, creating sharp edges. A crevice tool will reach into tight seams, around seat beads, and under seats. Vigorous back-and-forth movement on carpet brings up sand and grit. An old hairbrush works well to break loose pine needles and other items stuck in carpet nap. A soft brush attachment on the vacuum works well for dusty dashboards, armrests, and consoles.

Pull out all of the floor mats and vacuum them outside of the vehicle. Move the front seat to its most forward position. Start vacuuming the back seat first. Use your free hand to separate the cushion from the seat back. You'll be surprised by the amount of debris that accumulates in this space. Spread the pleats to get at tiny bits of grit. Use the crevice tool around beads on the seat. You should be able to reach the entire back seat from one side of the car. If not, do one half and then move to the carpet in that area.

With the back seat vacuumed and the front seat still in its most forward position, reach under the front seat to vacuum as much carpet as possible. Briskly agitate the carpet with the crevice tool to bring up embedded dirt. Sometimes you can tap carpet with the palm of your hand. This will cause grit to pop up out of the nap and rest on top of the carpet, making for easy removal. Black carpet requires the most patience. It seems that the more you vacuum, the more grit appears. Hang in there, as the end results will be well worth the effort.

After the back seat and carpet have been vacuumed, move the front seat as far back as it will go. Vacuum the front seats and then the carpet. Use the same thorough approach as you did in the back. Vacuum the wells in the armrests on door panels inside the console and

1. Pull out the floor mats before starting your vacuuming project. They will be cleaned outside of the vehicle. Moving the front seat to its most forward position will give you the most available working room in the back seat. Start vacuuming at the highest spot in the rear seat area. This way, dust and lint will fall on a surface that has yet to be vacuumed. Use the crevice attachment for powerful suction in concentrated spots, and employ other attachments as necessary for optimal cleaning.

Detailing Techniques

glove box. If the dashboard is dusty, use a soft brush attachment to clean vents, gauges, and grooves. Don't forget the area and carpeting surrounding the center console.

Some consoles include cloth pads. A toothbrush used in front of the crevice tool works well to loosen dirt along the seam between the cloth and plastic. A toothbrush also helps to break up dirt accumulations in glove boxes, on foot pedals, and along grooves in steering wheels and consoles.

Inspect the entire vehicle interior for any dust or loose debris that might be easily removed with the vacuum. Check the rear deck next to the back window, headliner, seat pockets, ashtrays, under the headrests, and so on.

2. Vacuum the seat back and work down toward the cushion. A wide attachment will work quickly to remove most loose dust, lint, and other debris. Use your hand to spread pleats apart and open up areas next to seams and beads. The crevice tool works best to remove unwanted material from these tight spaces. If you come across a stubborn spot of caked dirt, agitate the crevice tool back and forth on it to break it loose.

3. After the back seat has been thoroughly vacuumed, look at the back of the front seat. Run the vacuum over the back of it to remove dust and lint. Don't forget to raise the headrests and vacuum under them.

4. Vacuum the carpet with a wide attachment first to pick up the big stuff. Pat the carpet with your hand to pop up loose grit that may be imbedded in the carpet fibers. Use a toothbrush to dislodge debris lingering in tight spots next to seat brackets and along edges. Hold the crevice tool close to the toothbrush in order to pick up the material that breaks free.

5. The crevice tool attachment also works well to reach inside tight spaces, like this area at the end of the rear seat cushion, next to the door opening. After you have completely vacuumed the entire interior, plan a few minutes to vacuum the trunk. Trunk spaces are frequently overlooked during car-cleaning exercises. It does not take long to vacuum the space and wipe it down with a damp towel.

Interior Detailing 41

TECHNIQUE 10 CLEANING THE INTERIOR AND DASH

Time: 1/2 to 2 hours

Tools: Towels, cleaner, toothbrush, cotton swabs

Talent: ★↙

Tab: $10–$15

Tip: Start in the back and work your way out of the vehicle

Gain: A more comfortable and pleasing ride

Complementary project: Clean the windows

With the front seat moved as far forward as it will go, you will have access to the front seat brackets for cleaning. This will also maximize the amount of room you have to work in the back seat area.

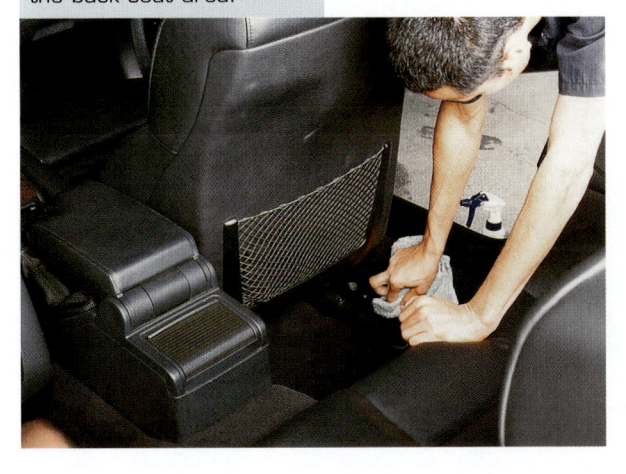

Rear Seat Area

When you are cleaning, the easiest way to reach most areas in the rear seating compartment is to sit on the seat. If your shoes are wet or dirty, place a towel on the floor to protect the carpet. A clean, damp wash mitt or small towel will remove dust and lint. Excess water streaks from the mitt are dried with a clean towel.

Start with the headliner above the seat. Then work on the far side panel, the rear deck area, the near side panel, and so on. Use a planned pattern, cleaning from the middle of the car toward the outside. Work your way out of the car by wiping the seat last. You can also easily reach the back side of the front seats. This will help you to maintain complete coverage and prevent the need to lean over a previously cleaned part to reach a dirty one.

For interiors in need of more than dusting, use a multipurpose liquid cleaner. Keep in mind that each time you use a cleaner, you will be removing some of the protection provided by the latest application of dressing or conditioner. Plan to reapply the appropriate conditioner after cleaning.

The liquid cleaner you use should be mixed with water according to instructions on the label. A spray bottle makes for the easiest application. Before spraying the headliner and side panels, fold a clean towel in quarters. Dampen one side with the cleaner and wipe the area to be cleaned; use a dry side to remove residue. Spray directly on an area when it is so encrusted with grime buildup that a simple rub with the towel won't work.

Light bursts of spray on a seam or groove will help loosen dirt. Use a toothbrush as necessary. Place a towel below the work area to catch runoff. Take your time and aim the bottle accurately. Overspray will have to be cleaned, making more work for you. You can place the toothbrush on a towel and dampen the bristles with cleaner before using it to scrub a spot, minimizing overspray and runoff. A toothbrush also works well to clean the rings on window cranks.

Your systematic approach should include removing and cleaning rear ashtrays. Small ones are easiest to clean using your fingers. Put on a cotton glove and use it as your custom-fit cleaning cloth. You can try using a toothbrush, but the configuration of smaller ashtray styles doesn't allow much working room. For those ashtrays that suffer from heavy encrustations, soak them in a wash bucket after rinsing with a hose or in a sink. Once they are dried, paint them to look new.

Be sure to closely examine lower parts of the interior compartment. Scuff marks from shoes are common along side panels next to doorjambs and the bottom sections of the front seat. Marks on vinyl and plastic are removed by rubbing with the cleaner-dampened towel. For those

42 Detailing Techniques

marks unconquered by the all-purpose cleaner, try a name-brand spot remover. The chemicals inside these powerful cleaners work well, as long as you follow the directions on the label.

In essence, you should have touched every square inch of the back seat area before you move on to the front. Have you checked deep between the seat back and cushion, along the outer edge of the seat next to the side panel, and under folding rear seats? Have chrome strips been polished, and has residue been removed from the grooves of Phillips-head screws? How about lint left behind from the cleaning cloth on the rear deck and carpet?

Unless you plan to shampoo the cloth seat and carpet, the back seat area should be clean by this point. Judicious use of vinyl dressing will be covered in a later project.

Front Seat Area

Start with the headliner. Clean it as you did from the back seat. Clean the sun visors and the brackets holding them in place. Clean the back of the rearview mirror, a commonly forgotten item. Use the toothbrush as necessary to clean around trim and to remove buildup in the exposed slots of Phillips-head screws. Seatbelt brackets for shoulder straps should be examined and cleaned also.

Metal dashboards, common in 1950s autos, are polished and waxed. Meguiar's One-Step Car Cleaner Wax is very well suited for this, since it doesn't leave a dry powdery residue behind. Use a small sponge for application. The square edges and small size make for easy maneuvering around gauges and trim. Be sure the surface is clean before polishing and waxing. Use spray cleaner and a towel to remove dust and dirt, and a toothbrush along trim and screw slots. Remove residual moisture with a dry fold of the towel.

Once a year, and especially for neglected vinyl dashboards, you might consider soaking a clean vinyl dashboard with dressing and using a clean paintbrush to work dressing (like Armor All or Meguiar's Number 40) into crevices and seams. Let it soak in for a couple of hours or even a day. Then wipe the vinyl with a cloth dampened with a mild cleaner to remove the superhigh gloss and slippery feel. Such an application helps the dressing/conditioner soak into the material deeper to last longer and keep future applications to a minimum.

Some auto enthusiasts and detailers may disagree on soaking vinyl with dressing. They believe that dressing is too often used to cover up inadequate cleaning jobs, allowing dirt to build up even more. Their preferred method relies on thorough cleaning with soap and water or any of the vinyl cleaners on the market. This is followed with a very light application of dressing/conditioner.

Basic dashboard cleaning is simple yet time consuming. Lots of grooves and slots along the face of a dash frequently collect dust. Knobs get dirty and sticky, and the newer styles of steering wheels may present a detailer with bumpy textures and grooves to clean.

Cotton swabs work well for grooves, and damp ones will collect and retain dust much better than dry ones. Dip them in water and squeeze with your fingers to remove excess. To help remove dirt, dip them in soapy water or lightly spray them with cleaner. Use swabs in vents, too, and anywhere else you deem practical.

Dust in the tiny holes on the outside of sound-system speakers is removed with the brush attachment on the end of a vacuum cleaner hose. Sometimes a dry paintbrush is handy to help loosen stubborn dust. Speakers are delicate, and you must use caution when cleaning. If the cover cannot be cleaned with

Above Left: Dirty spots on cloth material are easily cleaned with use of a cleaner designed for fabrics. A number of different brands are available at auto parts stores, supermarkets, and some variety stores. Notice the brush-like cap on this product used to work in the cleaner on tough spots. Be gentle with headliners. Too much aggressive scrubbing can cause them to loosen and eventually sag. It is best to employ light, even pressure. A clean towel is used to pick up most of the cleaner lather. Again, use light pressure rather than heavy force.

Above Right: A wet/dry vacuum is used to remove lingering hints of cleaner lather and loosened dirt from the fabric. Along with helping to remove dirt, a wet/dry vacuum helps cleaned fabric to dry quicker.

As you work around the interior, remember that a dry paintbrush will quickly whisk away dirt and debris from tight spots. A paintbrush is employed here to remove debris from the windshield area next to the dashboard.

Above Left: "Dust Blaster" is a product commonly used to clean computer keyboards. Basically, it is cold air in a can. It works great for cleaning dash vents. The air coming out of the can is super cold, so short bursts work best.

Above Right: A cotton swab lightly dampened with cleaner makes quick work of dusting dashboard heater vents. Cotton swabs are also useful in removing more stubborn spots of dirt buildup. It may require the use of three or four cotton swabs to finally loosen and remove stubborn spots of dirt or grime buildup.

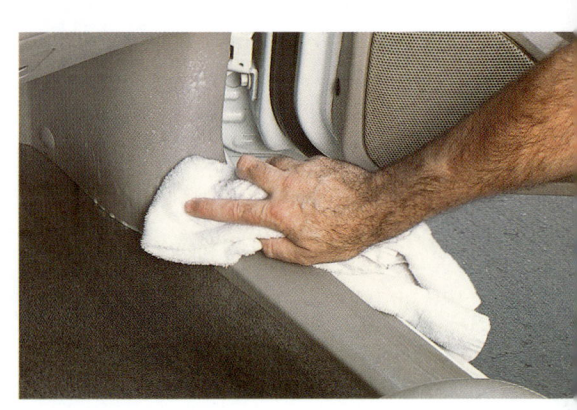

Below Left: A paintbrush is agitated around controls located on a center console panel. Power window and door lock buttons are common spots where dirt and grime will accumulate. Hold the end of a vacuum cleaner hose next to the paintbrush to pick up the dust that is dislodged. If stubborn dirt buildup is a problem, spray just a hint of cleaner on the paintbrush bristles to help them more quickly loosen up and whisk away dirt. Once again, Dust Blaster also works well to blow away dust and dirt from the exceptionally tight space around a door panel power window control.

Below Right: As you work your way out of the interior, be sure to look down to notice scuffs on any of the lower interior panels. Most scuff marks are easy to remove with an all-purpose cleaner on a towel or cleaning cloth.

Interior Detailing 43

a vacuum and dry paintbrush, consider removing it and cleaning it outside of the car with mild soap and water and a toothbrush. Be sure it is completely dry before reinstalling.

To clean sticky steering wheels and knobs, use cleaner on a towel and a toothbrush. Fold the towel as needed to reach hard-to-get areas. Use the edge of the towel along wide grooves on the center of the steering wheel. A toothbrush and cleaner are used to scrub dirty materials that have unusual textures.

The handle of a toothbrush placed inside a towel may be used to reach deep into dusty vents without fear of scratching. Some detailers like to spray vents with liquid cleaner first, then remove dirt with the toothbrush-in-a-towel method. The liquid breaks dirt loose, and the towel picks up residue. Use a cotton swab to gather excess along the sides.

Clean the glove box and console with a towel, toothbrush, and cotton swab. A paintbrush is handy to reach in grooves and deep glove box corners and storage wells on consoles. Be imaginative. Use the tools at your disposal to clean whatever needs to be cleaned. Always use the simplest and mildest methods first, before graduating to more potent means.

Seatbelts

Seatbelts must not be overlooked. Pull them out of their cases to check for dirt and proper rewinding. Use the crevice tool on the vacuum cleaner to remove dust and dirt from the case. If the return mechanism is squeaky or unusually slow, clean and then lubricate it with a very light coat of WD-40.

Clean seatbelts with soap and water, using a toothbrush or plastic scrub brush as needed. Lay a towel on the seat and clean the belt on top of it to prevent runoff residue from getting on fabric or carpet. Check the vehicle owner's manual if you question proper soap use. Allow the belt to dry before returning it to its case. To keep seatbelts stretched while drying, tie one end of a string to the end of the belt and the other end to the steering wheel, door handle, sun visor bracket, or another fixed point.

Dashboard

Clear lenses over gauges are cleaned with water or glass cleaner. Removing dust from the inside of gauge faces requires dashboard dismantling for gauge removal. Then the lens can be separated from the gauge itself. Unless you know what you are doing, think about seeking advice from a knowledgeable friend or professional before getting started.

If a lens is scratched, try polishing. Use a name-brand glass polish for glass; for plastic lenses, use the same polish that is made for clear plastic windows on ragtops. These products are manufactured as non-abrasive cleaners and polishes for plastic, Plexiglas, window tint film, porcelain, and other hard surfaces. If a glass or plastic lens is scratched and you are determined to fix it, try this mild method first while the gauge is still in place in the dash.

TECHNIQUE 11 DOORJAMBS AND PANELS

Time: 1/2 to 2 hours

Tools: Bucket, cleaner, mitt, towels, brushes, shampoo, and wet/dry vacuum for cloth panels

Talent: ★↗

Tab: $5–$15

Tip: Protect the interior from cleaner residue with large towels

Gain: Assurance that door drains are clear

Complementary project: Lubricate door hinges, latches, and key locks

Doorjambs

Use a damp wash mitt to pick up dust around door edges and the door-opening perimeter along the A- and B-pillars, roof, and rocker panel. Clean both the front and rear doorjambs, as well as the top and bottom door edges. If a damp mitt can't do the trick, try a towel dampened with cleaner. Use a paintbrush dipped in a wash bucket of soap and water to reach difficult spots. If need be, spray an all-purpose cleaner like Simple Green on tough dirt and grime deposits. Employ a soft brush or toothbrush to dislodge buildup in seams and along hinge edges. Wipe off residue with a clean soft cloth or towel.

Dry door frames with a clean towel. Painted surfaces look great when they are wet, but once they dry, missed spots will stand out. Clean and dry again as needed.

Most automobile doors are equipped with drains. They are nothing more than holes along the door frame's bottom edges. You have to get your head below the door and look up to see them; you should look at and thoroughly clean these drains a few times a year. Should they plug up, you run the risk of giving rust a start in the bottom of the door. Clean plugged drains with a pipe cleaner and vacuum.

Moldings around doors serve very useful purposes. Clean them with a towel and look for sections that may be coming loose. Condition them with an all-purpose dressing and reattach loose sections with an appropriate molding adhesive. These are found at autobody paint and supply stores and most auto parts outlets.

Check your vehicle's owner's manual to see if hinges and latching mechanisms require special lubrication. If not, it is a good idea to lubricate them after a complete cleaning using a spray product like WD-40. Sprayed through its plastic tube, WD-40 works well on hinges and inside keyholes on doors. An occasional shot helps keep hinges from squeaking and key mechanisms working freely to allow easy key turning.

Service stickers on doorjambs are unsightly to most car enthusiasts. Use your fingernail to gently remove them. If at

Doorjambs and the surrounding door opening perimeter are easy to clean with a damp towel. Plan to do this after drying your vehicle when you've finished washing it. Areas around the door hinges may require some cleaner and agitation with a paintbrush.

Above Left: A toothbrush is used to dislodge an accumulation of dirt at the door hinge area. Since hinges are lubricated, it is easy to understand why dirt seems to accumulate around them more than at other spots on doors.

Above Right: Paintbrushes work well to break loose dirt and other buildup around door edges. By themselves, they will break up and whisk away dry pockets of dirt. Lightly spray the bristles with an all-purpose cleaner to help them loosen up dirt and grime accumulations.

any time you think you are about to damage paint, stop. Use a sticker and decal adhesive remover to loosen glue and then peel as necessary. Place some remover on a rag and use it to wipe away adhesive residue.

Doorjambs may be polished and waxed. Take your time and concentrate on applying polish and wax just exactly where you want it. Use a rectangular sponge as an applicator. Its straight sides work great for cutting an edge along vinyl and rubber trim and molding.

Door Panels

Door panels are cleaned much the same way as other vinyl or cloth interior accessories. Scrub vinyl, if necessary, to remove shoe scuffs and other such marks. You might also try using name-brand spot removers especially made for removing marks from vinyl and similar surfaces. Use caution around imitation plastic chrome strips. Vigorous scrubbing will cause their thin plastic imitation chrome film to peel away from its plastic trim base. Thoroughly clean door handles and window cranks, using a toothbrush as needed. Be sure to clean and dry the wells on armrests and the grooves on door speakers.

Neglected vinyl door panels might be easiest and quickest to clean out on a driveway. Open the vehicle door and cover the seat and carpet with large towels. Mix up a nice sudsy solution of all-purpose cleaner and water. Dip the mitt into the soapy solution and use it first to go over the entire door panel. Work on tough dirt and grime spots with a plastic-bristled brush loaded with soapsuds. Rinse with a gentle flow of water from a garden hose. Repeat the process as necessary to get rid of dirt, grime, scuff marks, and other related blemishes. Note: Refrain from this aggressive cleaning method if door panels are equipped with electronic sound system components or other moisture-sensitive equipment. In those cases, use towels dampened with cleaner and follow up with a water-moistened towel.

Cloth panels should be vacuumed thoroughly before you attempt more aggressive cleaning tactics. Once all of the loose stuff has been removed and you find that more cleaning strength is necessary, consider using an upholstery spot remover or spray-on shampoo. Follow label directions and then remove the dry residue with a vacuum.

Severely soiled cloth panels may be cleaned with liquid shampoo and water and then rinsed with a gentle flow of water from a garden hose. Spray the panel with a shampoo solution poured into a squirt bottle. Then dip a plastic-bristled brush into your wash bucket of shampoo and start scrubbing to bring up a good lather. Rinse with a light stream of water from the garden hose, using caution to prevent water from running or falling into the interior compartment.

You must have a wet/dry vacuum ready to assist in removing shampoo and water from the cloth material. Be certain the crevice tool on the vacuum is squeaky clean before touching it to the panel surface. Apply light pressure on the crevice tool as you slide it down

Detailing Techniques

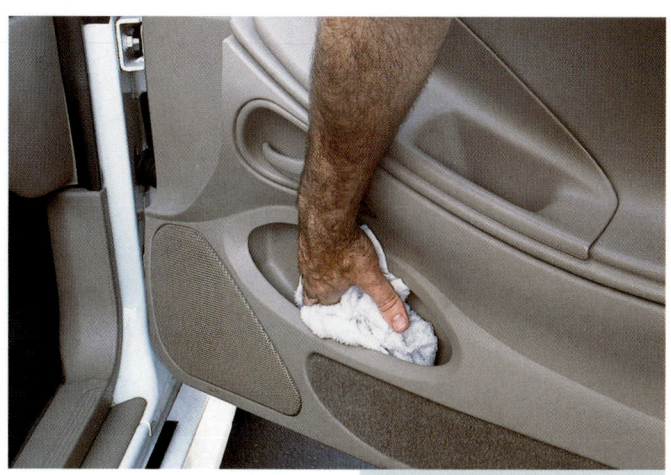

the panel to help force out liquid from the cushiony fabric. Go over the panel three or four times to ensure you have removed as much moisture as possible. Leave the door(s) open in the sunlight for a couple of hours to aid the drying process. You can also direct the breeze from a fan onto the panels to help speed drying.

Never simply close vehicle doors with the windows up and expect damp cloth panels to dry. Moisture from them will quickly evaporate into the interior to cause overall interior moisture and potential mildew problems. If you do, you'll likely notice the windows fog up within a matter of minutes. At that, even if you are convinced that the door panels are dry, plan to park the vehicle inside a garage with the windows down overnight. This should ensure that all of the moisture has evaporated and the cloth door panels are completely dry.

Above Left: Every door is equipped with door drains. This is how water escapes from inside doors after it seeps past window moldings during a rainstorm or after a wash. It is vitally important to keep these drains open. Plugged up, they will allow water to settle at the lower door sections and give rust a place to start.

Above Right: Door panels are dusted off with a damp towel. Those with dirt smudges may require the use of an all-purpose cleaner. Don't forget to clean the bottoms of pockets, handgrips, and other wells.

Really dirty door panels may be scrubbed with an appropriate cleaner and a scrub brush. Vinyl door panels are dried with a clean towel. Cloth panels will require the use of a wet/dry vacuum to assist in the removal of accumulated water and shampoo suds.

Interior Detailing

TECHNIQUE 12 VINYL AND RUBBER DRESSING

Time: 1/2 to 1 hour

Tools: Dressing, clean soft cloth, clean small soft-bristled brush or paintbrush

Talent: ★↗

Tab: $5–$15

Tip: Spray dressing onto a cloth and then apply to surfaces

Gain: Vinyl and rubber preservation

Complementary project: Apply a fresh or scented air freshener under the front seat

Vinyl/rubber protectant (dressing) may be applied in two ways: spraying directly onto the material or spraying it onto a cloth first and wiping it on. It is generally best to spray a cloth and wipe dressing onto dashboards to eliminate overspray on gauges and other dashboard accessories.

Multipurpose vinyl dressing like Armor All, Meguiar's Number 40, and Mothers Preserves will bring new life to faded interior vinyl. In the case of dressing, too much is not too good. You have likely noticed the slippery feel of vinyl seats conditioned with too much dressing. You need a seatbelt just to stay in one spot. It is best to apply a thin dressing coat rather than a heavy dose.

An exception may be interiors that have been neglected for extensive periods of time. On those, apply a heavy spray and work in dressing with a soft brush. Let stand for a while and then wipe off the excess with a clean towel. A few hours or even a day later, gently wipe the dressed parts with a towel dampened with a mild mix of all-purpose cleaner and water. This should remove excess dressing and get rid of the slippery feel of the material, while leaving the vinyl with a rich and rejuvenated appearance.

Auto enthusiasts generally have one towel set aside just for dressing use. They spray vinyl first and buff off the excess with that towel. The towel seldom gets dirty because it is used exclusively for applying dressing onto clean surfaces. In tight areas, use the towel by itself. In most cases, this towel is so impregnated with dressing that you won't have to spray any on the surface; residual dressing on the towel takes care of everything.

Most of the cosmetic car care product manufacturers offer dressing products. They work well when applied according to directions. Although you might find that serious auto enthusiasts have brand preferences, you should try different products until you find the one that works best for you.

For general applications on parts other than seats, spray dressing on a small clean towel. Using a folded edge of the towel or your fingers, guide the application carefully. Only vinyl and rubber need to be dressed. Dressing on plastic and metal parts does not look good and is unnecessary. Start in the middle of the back seat area and work your way out of the car, just as you did when cleaning. For hard-to-reach spots, place the handle of a toothbrush inside the towel and use it as an extension. As the towel dries, spray a little more dressing onto it.

Vinyl dashboards present unique problems during dressing application. Many nooks, crannies, and obstacles are

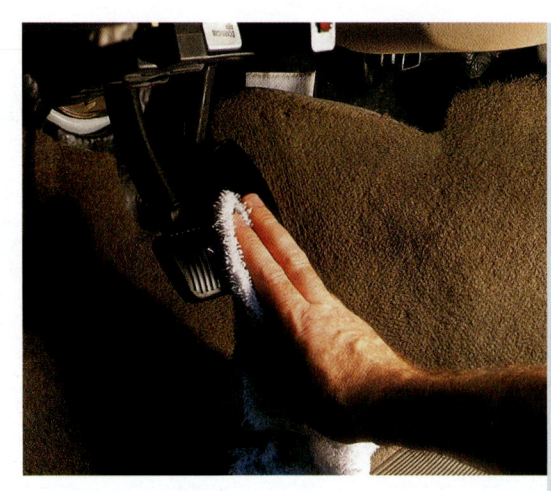

48 Detailing Techniques

Far Left: Meguiar's Self-Saturating Detailing Swabs are very handy in the application of dressing to tight spaces along dashboards, consoles, and other interior features. They will quickly reach in where your finger in a cloth will not.

Left: Foot pedal rubbers should be lightly dressed. Be certain to buff off all excess to ensure they are not slippery. If necessary, allow dressing to soak into the foot rubbers and then wash them off with soap and water to remove the slippery sheen and leave them looking new.

in the way. Use your fingers as guides. On areas most tightly confined, spray dressing directly onto a small towel and wipe onto the surface. This lessens overspray problems and allows you the chance to put dressing exactly where you want it.

In really tight spots around dashboards and consoles, try using dressing-filled swabs made by Meguiar's. The swabs' cotton ends are attached to hollow strawlike tubes filled with dressing. Dressing flows from the tubes to the cotton tips for smooth and controlled application. Along with swabs, Meguiar's also offers handy dressing wipes. These dressing-saturated, light fabric "towelettes" are easy to maneuver, and they work great for quick interior vinyl and rubber spiffs.

Foot pedals and rubber mats may also be dressed. For these parts, apply only a very thin coat and be absolutely certain you remove all the excess. Seriously consider wiping off these parts with a towel dampened with cleaner afterward to ensure they are not too slippery. If too much dressing is applied to parts like these, the bottoms of your shoes may become slippery enough to cause your feet to slip off the pedals, which could possibly result in an accident.

Rubber moldings are wiped with a cloth dampened with dressing. This makes them look new. The same holds true for all rubber and vinyl interior parts. Using the same pattern you did when cleaning, assure yourself of complete coverage. Afterward, park the vehicle in the sunlight and inspect the interior for any missed spots.

Give door moldings a light treatment of dressing to keep them looking new and remaining pliable. You may have to use a toothbrush to work dressing into tight spaces along the outer door edge.

Interior Detailing

TECHNIQUE 13 VINYL SEATS

Time: 1 to 2 hours

Tools: Cleaner, towels, scrub brush, toothbrush, vinyl spot remover

Talent: ★✦

Tab: $5–$25

Tip: Vacuum first and concentrate around beads, seams, and pleats

Gain: Clean seats will not soil clothes

Complementary project: Clean the seatbelts

Cleaning

Since vinyl seats won't absorb water, you can scrub dirty ones with a plastic-bristled brush and soap and water. This procedure is not necessary on lightly soiled seats, which are easily cleaned with a towel and an all-purpose or name brand vinyl cleaner.

Spray liquid cleaner on a back seat section, starting at the top of the backrest, and work down. Use the brush in a circular pattern to break loose embedded dirt from the grain. Then reverse the brush direction for best coverage. Use the brush along seams and pleats, adding cleaner from a spray bottle as necessary to maintain a foamy base. Place towels everywhere you want to prevent overspray or splattering.

Dry with a towel. You may also use a wet/dry vacuum to remove suds and moisture, followed by towel drying. The crevice tool works well to remove moisture and dirt caught in seams and along beads. Be sure to dry lingering residue left on armrests, side panels, chrome strips, and deep creases.

Scrub one seat section at a time. For example, do the passenger's side of the backrest of the rear seat, then that half of the bottom cushion. Move on to the other side and do the same. Then move the front seat to its most rearward position and do the driver's side, then the passenger's. This pattern assures good coverage and prevents soap from drying on the vinyl.

Stubborn stains left behind will have to be removed with a vinyl spot remover. Most of these products advertise similar qualities and will do the job of stain removal very well. Be very cautious about using solvent, paint thinner, or lacquer thinner. These harsh chemicals may remove stains but could also quickly damage vinyl. Try repeated applications of vinyl spot remover on vinyl before contemplating the use of any harsh solvents or thinners.

Exceptionally neglected and filthy seats may be easiest to clean and "disinfect" if they are first removed from the vehicle. Out in the open, you will enjoy unfettered access to all parts of the seats for optimum cleaning maneuvers. Refrain from thoughts of rinsing

Vinyl seats are initially cleaned with an all-purpose cleaner and a clean cloth. Stubborn spots are removed with vinyl cleaner and a toothbrush or scrub brush. Be sure to wipe away residue with a clean cloth dampened with water.

Detailing Techniques

Above Left: After vinyl seats are thoroughly cleaned, you need to replenish the oils and lubricants removed from the material during the cleaning endeavor. Meguiar's Vinyl/Rubber Protectant and Mothers Preserves work well.

Above Right: Once a vinyl dressing, or protectant, has been applied, you must use a clean cloth to wipe off (buff) the excess. Too much dressing is not good. Seats will be extra slippery and uncomfortable. Buff the material until you arrive at a finish that is crisp but not slippery.

seats with a garden hose, though. Too much water will surely filter into the cushion and padding and cause interior moisture problems later on. If you find that you have to employ a lot of scrubbing with loads of cleaner, wipe off the seats with a water-dampened towel before drying. This will remove excess cleaner residue to leave the vinyl feeling clean and smooth.

Dressing

Freshly cleaned vinyl should be treated to a light coat of vinyl protectant, such as Armor All, Meguiar's Number 40, or Mothers Preserves. These conditioners will replenish some of the oils removed by cleaning products to help make the material look and feel like new. They also go a long way toward helping to preserve the material and protect it against wear and the adverse effects of the sun's ultraviolet rays.

Spray dressing directly onto the seats and work it in with a clean soft cloth or towel. Use a soft-bristled brush to work the dressing into areas around pleats and seams. Wipe with a clean towel to remove the excess.

Control the dressing spray with a good aim to prevent overspray onto interior parts. As you notice that the dressing appears to be drying out, spray some more on. Your goal is to end up with all sections of the seats looking alike in luster tend richness. Be certain to remove all hints of dressing excess. If you see that the material has become too slippery or too glossy for your taste, wipe down the seats with a clean towel lightly dampened with a mild cleaner.

Vinyl Seat Repairs

Most auto parts and autobody paint and supply stores carry vinyl repair kits. They generally sell for around $15. These can be used to repair small tears, punctures, and burn marks. You will have to read labels to determine which vinyl colors are included in the kits. If the vinyl seats in your car have been dyed, or the material is of a custom sort, you might have to get help from a professional upholstery shop. Major tears, split seams, and other more serious vinyl seat problems need to be repaired at an auto upholstery shop. Check around to determine which shop offers the best price.

Most auto parts stores sell vinyl repair kits for around $15. These work well for minor tears and blemishes on vinyl upholstery. The kits include color-matching pastes, textured backing material, and a heat iron.

TECHNIQUE 14 CLOTH SEATS

Time: 1 to 3 hours

Tools: Vacuum, upholstery cleaner of choice, brush, towels

Talent: ★★★

Tab: $10–$20

Tip: Don't plan on driving a vehicle with wet, shampooed seats until the next day

Gain: Seats that look new and won't soil clothing

Complementary project: Apply a cloth upholstery protectant

Soiled cloth seats must be shampooed. Some detailers have enjoyed good results using shampoo products designed to dry after application, with residue being picked up with a regular vacuum cleaner. These work well for lightly soiled upholstery. Be sure to follow label instructions closely.

Dry Cleaning

The fabrics used for automotive cloth seats are similar to some fabrics used for household furnishings. Many detailers have enjoyed good results cleaning automotive seats with upholstery shampoo purchased at a local supermarket. If you decide to shampoo, make sure you have a wet/dry vacuum cleaner available to remove most of the water and shampoo suds from the upholstery. This will also help speed up drying and will help prevent mildew problems.

Most of the automotive fabric upholstery cleaners found in auto parts stores work well, and they are easy to apply and remove. The same holds true for most fabric cleaners found in supermarkets and variety stores. It is tough to recommend one over the other, as auto enthusiasts have had luck using them all. Your best bet might be to stay with those products specifically designed for automotive upholstery applications and to ensure you follow label directions.

If you don't have access to a wet/dry vacuum cleaner and prefer not to rent or use one at a self-serve car wash, you will have to use an upholstery shampoo product designed for removal with a dry vacuum cleaner. Auto parts stores should carry an assortment of brands. Be sure to follow the instructions on the label.

To remove stubborn stains on cloth seats without wet shampooing, try using a dry spray-on upholstery stain remover. You can find them at auto parts stores and supermarkets. These products spray on wet, with the outer edges drying first to form a white ring around the stain. The rest of the spot remover eventually dries to a white powder, too. Use a clean plastic-bristled brush to whisk away powder and stains, and then vacuum up remaining residue.

Thick debris stuck to cloth seats can be scraped with a dull knife and then blotted with a rag dabbed in cleaner. Scrub the remaining residue with a toothbrush and shampoo. Use a wet/dry vacuum cleaner in the direction of the fabric grain. Speed drying with a hair blow dryer.

Wet Shampoo

Plan to start at the top of the rear seat backrest and work down. Fill a spray bottle and a small bucket with shampoo mixed with water according to instructions. Spray a small section of seat first; then dip the clean plastic brush in the bucket and start scrubbing. Shampoo agitated by a scrub brush on top of upholstery should result in a rich foam blanket. Scrub in a circular direction,

Detailing Techniques

reversing it occasionally for added cleaning power.

Vacuum each section immediately after scrubbing. Use a clean crevice tool at the end of your vacuum hose. Apply light pressure to the crevice tool as a means to help force shampoo and moisture out of the thick seating upholstery. This will prevent shampoo from drying on the surface and will remove most of the moisture before it penetrates too deeply into the cushion and padding. Use a clean towel to dry other things, like chrome strips and seatbelts. Once the entire back seat has been shampooed and vacuumed dry, move on to the front seat to repeat the process. Inspect your work and go over those spots that did not clean up on the first pass.

When the entire shampoo job is done, park the vehicle outside with the windows open to assist the seats in drying. During winter months, open the doors and place a fan at an opening, allowing air to blow over the seats and hasten the drying process. Another way to help speed the drying process is to park your vehicle outside, turn on the engine, and let it idle. Crack all of the windows open about an inch and turn on the car's heater full blast. Let the car run for an hour or so, checking its temperature occasionally to ensure the engine is not overheating.

Above Left: Wet-type upholstery shampoos are sold in auto parts stores and most supermarkets. Mix up a solution according to directions and pour some in a spray bottle and some in a bucket. Spray the shampoo on the soiled area, dip a scrub brush into the bucket, and begin scrubbing.

Above Right: After a section of upholstery has been shampooed, use a wet/dry vacuum to pick up the lather and lingering moisture. This will not get the upholstery completely dry, but it will help to speed the drying process while picking up residual dirt and shampoo suds.

When you have completed the shampoo project, take a few minutes to wipe down surrounding seat parts and other spots where shampoo may have splashed.

TECHNIQUE 15 LEATHER SEATS

Time: 1 to 3 hours

Tools: Leather cleaner, leather conditioner, soft cloths, towels, soft scrub brush, toothbrush

Talent: ★↲

Tab: $10–$25

Tip: Always apply a leather conditioner after using leather cleaner

Gain: Preservation of the leather upholstery

Complementary project: Clean and apply appropriate conditioner to the dashboard

After a thorough vacuuming with the crevice tool, striving to get rid of the loose stuff along beads, seams and between pleats, wipe down leather seats with a clean water-dampened towel. Then dry with another clean towel. Never use ordinary all-purpose cleaning products or cleansers on leather. Rather, plan to use only those cleaning and conditioning products especially designed for leather. Such include those specific leather care products made by Meguiar's, Mothers, Eagle One, Lexol, Hyde Food, and so on. You can also find an assortment through The Eastwood Company.

Application of a leather cleaner product is simple. Just dab a small amount onto a damp sponge or clean soft cloth and scrub in an easy motion. Reapply as necessary for extra-dirty areas. For tougher spots, pour some leather cleaner onto the bristles of a toothbrush or other soft brush and scrub gently. It is best to go over stubborn spots a number of times with soft pressure rather than once with too much. If that doesn't work and the seat is still unusually filthy, try using saddle soap.

An easy pattern to follow is to push the front seat as far forward as it will go and then start cleaning at the top of the rear seat backrest and work down. Do the backrest first and then the seat cushion. You can start on one side and work toward the other, or start in the middle and work your way out the door. Place towels on the carpet to prevent unnecessary soiling if your shoes are not completely clean.

Once the rear seat has been cleaned, focus your efforts on the front. Move the front seat as far back as it will go. This should give you plenty of open working room. Start at the top of the backrest near the middle of the seat and work down. Clean that half of the backrest and then the seat cushion. Work your way out of the door, ensuring you have cleaned both sides of the seat cushion and along the side of the console.

When you have completed a seat section, use a clean water-moistened towel to remove resi-

Leather upholstery is special. Treat it by using only those products designed for leather care. Once the leather seats have been cleaned, be sure to treat them to a leather conditioner applied according to label instructions.

Detailing Techniques

due. Use your hand to spread the leather, stretching folds in the material to ensure complete coverage. Be sure to clean and dry under buttons and along seams, beads, and headrests.

After cleaning and checking that the material is dry, treat leather to an application of leather conditioner. This is applied in much the same manner as cleaner. Place a small amount on a clean cloth and massage it into the leather. Start at the top of the back seat and work downward, paying close attention to the pattern followed to ensure complete coverage. Remember to treat the back of the front seats, too. Once the back seat is done, move on to the front. If you feel as though too much conditioner has been applied, use a dry side of your application cloth to buff off the excess.

1. Cleaning leather seats is not much different from cleaning ordinary vinyl seats, except for the products employed. Spray or wipe leather cleaner onto the surface, doing just a small section at a time.

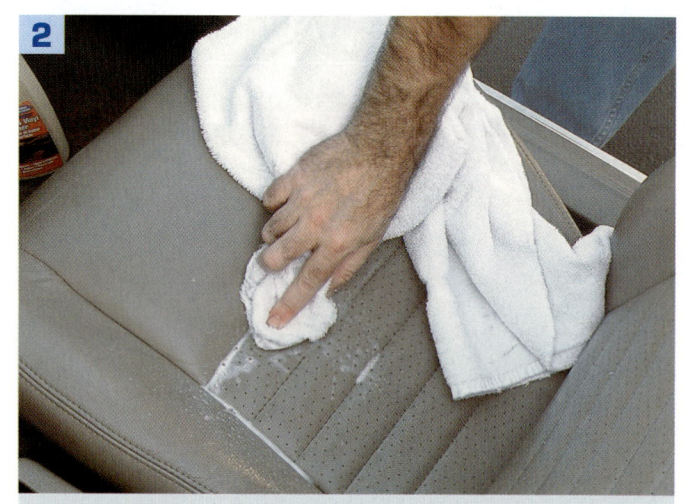

2. Use a clean cloth to work in the leather cleaner. Rub in a circular pattern, reversing direction to ensure complete coverage. Work your fingers into the pleated sections and areas around seams.

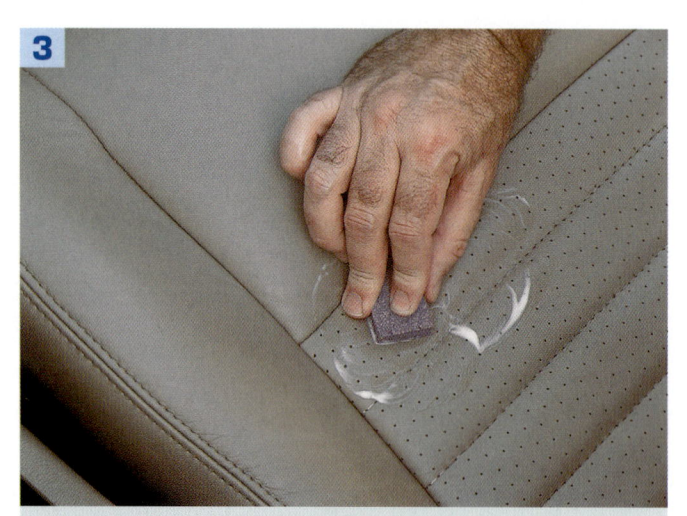

3. Use a small cloth or section of a mild scouring pad to clean dirty areas along seat seams. Make sure you have plenty of leather cleaner on the surface before scrubbing. Your free hand can be used to pull seams and pleats apart for best access to the area.

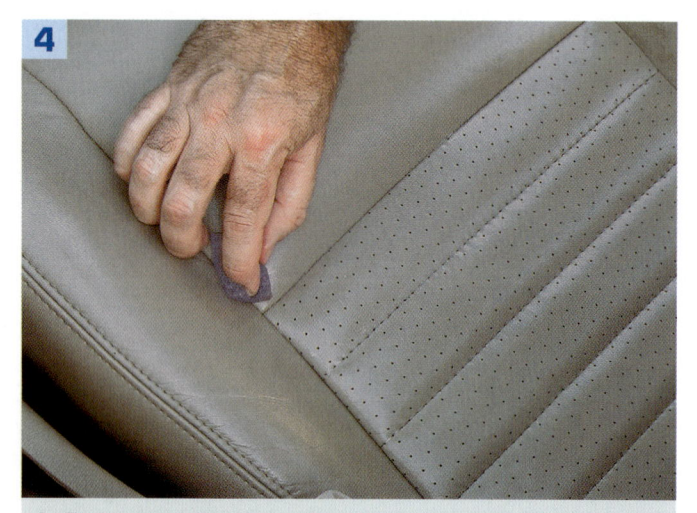

4. Stubborn dirt spots and stains may require the light use of a scouring pad. Be gentle. It is much better to employ a number of applications than to scrub once with a lot of pressure. Remember that leather is tough, but its finish can be damaged with aggressive scouring.

TECHNIQUE 16 SHAMPOOING CARPETS

Time: 1 to 3 hours

Tools: Wet/dry vacuum, bucket, spray bottle, carpet shampoo, scrub brush, towels

Talent: ★↘

Tab: $10

Tip: Make sure the carpets are 100 percent dry before rolling up the windows completely

Gain: New-looking carpet and a fresh smell throughout the interior

Complementary project: Clean and apply dressing or conditioner to the seats

A number of dry carpet cleaners are available at supermarkets, variety stores, and some auto parts houses. Most of these products are designed to be rubbed into carpet with a sponge. Once dried to a powdery residue, removal is accomplished with a regular vacuum cleaner.

With a wet/dry vacuum cleaner, you can shampoo carpets using any number of liquid carpet shampoo products. Supermarkets carry an assortment of brands, some with specific uses. You must read the labels to determine which will work best for you. A few detailers have experienced good results using Woolite as a carpet shampoo.

The process is rather simple. You will need a plastic-bristled brush, a spray bottle, a small bucket, some clean towels, and a wet/dry vacuum cleaner. Remove the floor mats and thoroughly vacuum the carpet to remove all loose debris and grit. Spend an adequate amount of time vacuuming to ensure an in-depth and complete job. It is much easier and more efficient to pick up loose, dry material than it is embedded wet stuff.

Start your shampoo efforts in the back seat area on one side of the car. Spray shampoo mix from the spray bottle on heavy stains and spots as needed. Next, agitate the brush in the small bucket to bring up foam and suds. It is not necessary to use a lot of water and shampoo with the brush; suds and foam work just as well and prevent carpet from becoming more soaked than necessary. Scrub the carpet in a circular pattern, reversing the direction occasionally. Bring up a good lather. Heavy stains should be sprayed again with shampoo and scrubbed for added cleaning power.

When a section has been scrubbed to your satisfaction, use the wet/dry vacuum with the crevice tool to remove suds, dirt, and moisture. Press the tool against the carpet while vacuuming. This will help to force water to the top and allow suction deeper into the nap. The vacuum cleaner will sound different when it is picking up water; you will easily notice the change in tone. After two to three passes, it will sound as though nothing is being taken in. Continue vacuuming for another four or five minutes. Although most of the shampoo and water has been removed, additional vacuuming will help to pull up small pockets of moisture that were missed.

Afterward, use a clean towel to wipe off splatter on seats, panels, and trim. Complete the back seat area first and then move the front seat back as far as it will go. Then start working on one side of the front seat carpet area. Once all of the carpet has completely dried, fluff the nap with your hand or a clean dry brush.

Shampooing is not often necessary. Simple stains are removed with dry spot removers. It is not good to soak carpets with water and shampoo, as moisture can remain in the padding for a long time, especially in cold, wet weather. The best time to shampoo is during warm weather, when you can park the clean car outside with the doors open to aid drying. For carpets that need it, try frequent maintenance with dry cleaners, saving wet shampooing for once a year.

Carpeted floor mats are cleaned in the same manner. Vacuum them completely,

Detailing Techniques

occasionally swatting them against a post or hitting them with the back of a foxtail broom to knock loose embedded grit. Spray them with shampoo, dip the brush into the bucket, and scrub away. Use the wet/dry vacuum to pull up dirt and shampoo residue, as well as moisture. Hang them over a fence in the sunshine to dry.

Carpets should not be dyed a different color. If carpets are faded, you can renew them with a light coat of dye the same tint as the original hue. Apply as instructed in the label directions, and then rub it in with a clean sponge to prevent a crust from forming on top of the carpet.

1. Dry-type carpet cleaners are available in auto parts stores, supermarkets, and some variety stores. They spray on wet and dry rather quickly. The benefit of using such products is the reduction in the amount of moisture brought into the automobile's interior. Once the dry-type carpet shampoo is sprayed on, use the scrub brush cap to work it into the carpet nap. This will help to loosen embedded dirt and grime. Employ a circular pattern and reverse it occasionally to ensure maximum cleaning strength.

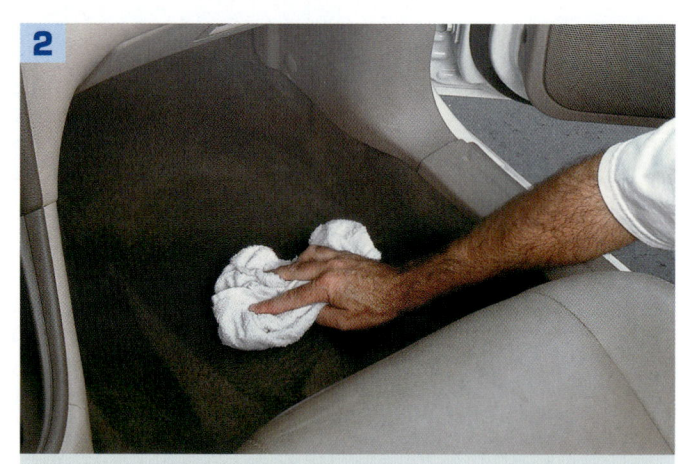

2. After you have completed scrubbing a section of carpet, use a clean towel to blot up moisture. You can use a wet/dry vacuum if one is available, but it is not necessary. Follow label directions and then let the vehicle sit in the sun with the windows open to help the carpet dry completely.

3. Wet-type carpet shampoo is mixed according to label instructions. It is then poured into a spray bottle and a small bucket. Spray carpet with shampoo first, concentrating on the dirtiest spots. Dip your scrub brush into the bucket and start scrubbing in a circular pattern, reversing direction occasionally to ensure complete coverage. As shampoo starts to dry out, add a little more from the spray bottle or the bucket.

4. Once an area has been successfully shampooed, use a wet/dry vacuum to pick up the shampoo suds and lingering moisture. Using a wet shampoo means that you will have brought a lot of moisture into the interior compartment. It is important that you vacuum the area for a few minutes to pick up as much moisture as possible. After vacuuming, lay a clean dry towel over the carpet and attempt to blot it dry. This will help to remove lingering moisture and help speed the overall drying process. After blotting the carpet, be sure to use a dry towel to wipe off exposed seat parts and other areas that may have been subjected to shampoo splatter.

Interior Detailing 57

TECHNIQUE 17 CLOTH PROTECTANTS

Time: 1/4 to 1/2 hour

Tools: Scotchgard or other protectant product, clean towels, vacuum

Talent: ★↗

Tab: $5–$10

Tip: Make sure the upholstery or carpet is clean and dry

Gain: Upholstery and carpet preservation and longevity

Complementary project: Clean the doorjambs and frames

After you have cleaned and shampooed the carpet and upholstery in your special vehicle, think about protecting it for the future. 3M makes Scotchgard for both carpet and upholstery materials. It is easy to spray on. Make sure you wipe off the excess, according to label instructions.

Fabric protectants are products designed to protect cloth upholstery against stains and light moisture absorption. Scotchgard is a familiar name and it works quite well. Similar products made by other companies are also available.

Most new car dealerships offer this type of upholstery and carpet protection for a fee. You can do the same job for much less. Apply cloth protectants only when upholstery is clean and dry. Spray the fabric in an orderly fashion and ensure you allow plenty of time for the material to dry before use. Be sure to spread pleats and maintain complete coverage. Always read label instructions and follow directions carefully.

Scotchgard is designed to help cloth resist stains and liquid absorption. It works well on cloth seats and carpet. Application is very easy, but you must remember that the fabric has to be colorfast and clean.

To apply, you simply spray it directly on the upholstery or carpet. Place towels strategically to protect windows and side panels from overspray. Use your free hand to stretch the fabric tight. Spray in a uniform pattern with a sweeping motion, 6 inches from the upholstery or carpet, with a speed that evenly wets the material. Overlap each spray by about an inch.

Plan on letting the protectant stand for at least three hours before sitting on the upholstery or placing your shoes on carpet. If a powdery white residue appears after drying, simply brush or vacuum it away. The powder is caused from applying too much protectant. On the other hand, if water droplets fail to bead up on the treated cloth, you must repeat the application.

This product is recommended for all types of fabric except leather, vinyl, and imitation suede. Applied correctly, it will help to prevent stains on your cloth seats and carpet. Its lasting quality is not uniform, as use and exposure to sunlight will lessen its overall durability. Once every couple of months, test its strength with just a few drops of water. If water soaks in, it is time for a new application.

As you would for cleaning seat upholstery, start with the back seat. Push the front seat forward as far as it will go to give you the most working room in the back. Begin in the middle of the seat at the top of the backrest and work down. Repeat the process for the other half of the back seat.

Start in the middle of the front seat at the top of the backrest and work down. Move to the other side and complete that half. Leave the doors open or the windows rolled down to assist drying.

TECHNIQUE 18 INTERIOR EXTRAS

Time: 1/2 to 2 hours

Tools: Clean cloth, Pledge, cotton swabs, air freshener, cleaner, scissors, window cleaner

Talent: ★

Tab: $5–$50

Tip: Inspect the interior with the vehicle parked in sunlight for optimum clarity

Gain: Fine-tuning the interior detail

Complementary project: Wash the exterior of the vehicle

After spending a lot of quality time detailing the inside of your favorite automobile, you should have developed a keen eye for perfection. You might even be able to go to a car show and note small flaws on some of the exhibitions. The more you clean an automobile interior, the more attentive you will become to the details. Before you know it, you will automatically clean all those little spots so commonly missed by less enthusiastic detailers.

To complete an interior detail, there are some final chores to consider. Line up all the vents on the dashboard so they point in the same direction. If seatbelts are not automatically retracted into holders, fold them neatly across the seat or fold them into a tight roll and gently tuck them into the space between the backrest and the cushion. Line up the rearview mirror so it is in its proper position and square with the top of the windshield. Replace worn floor mats with new ones, and be sure the straps for sheepskin seat covers are properly attached and not hanging loose.

Clean metal ashtrays should be painted with bright silver paint to look new. Clean plastic ashtrays can be painted with crystal clear lacquer to make them look unused. Foot pedal brackets may be carefully painted with semigloss black, using rags or towels to protect against overspray. Floor shifters might also be painted or their chrome surface polished. Plastic consoles can be polished with Pledge or other furniture polish—apply polish to a clean towel, wipe on, and buff off the excess.

Minor tears in upholstery should be repaired as soon as possible. The longer you wait, the worse tears will get. You can buy vinyl repair kits and do the job yourself, or go to an auto upholstery shop to get the job done. Sometimes, it might just be best to get entire seat sections replaced. As you inspect the carpet and upholstery, use a pair of scissors to trim loose threads or protruding carpet nap fibers.

Filthy ashtrays not only look bad, they smell bad, too. Be sure to clean them with soap and water. A toothbrush or small wire brush may be needed to get rid of the encrusted buildup. Spray clean and dry metal ashtrays with bright silver paint to make them look new. Spray plastic ashtrays with crystal clear lacquer paint to make them look unused.

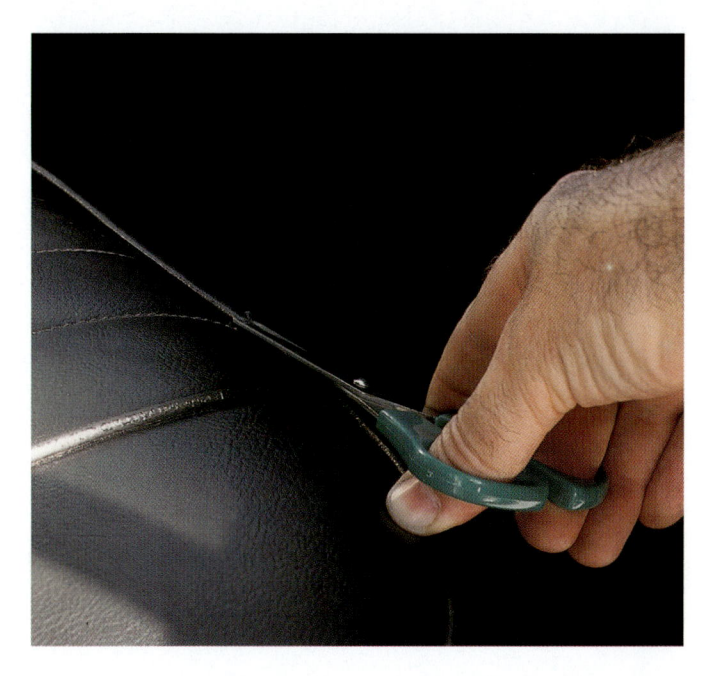

Above Left: In sunlight you will easily notice spots on the upholstery that were missed with either vinyl dressing or leather conditioner. Place a small amount of the appropriate conditioner on a cloth and wipe on until the entire seat looks uniform and crisp. Spots of dirt and light stains also may be addressed while your vehicle is parked in the sunlight. Just dab a small amount of cleaner on a cloth and wipe the spot away. Follow up with a light coat of conditioner.

Above Right: While inspecting the interior and looking for flaws, don't pass up the opportunity to trim loose threads from upholstery. Use a pair of sharp scissors instead of a knife to ensure a smooth trim without worry of unraveling. Use scissors to trim carpet nap that stands up out of place, too.

The little things make the difference between a good detail and a great one. Are any dead moths lying at the bottom of the dome light? Do sun visors complement each other by lining up in the same position? Are the bottoms of door pouches clean? Does the glove box look like you just threw a bunch of stuff in it, or does it look neat and tidy when you open the door? Are the windows perfectly clean inside and out? Subtle things like these will make an otherwise ordinary automobile stand out, looking special, unusually crisp, and clean.

Air Fresheners

Many car people do not use air fresheners inside their automobiles. Rather, they prefer the smell of clean. This "clean" smell can be maintained if you clean the inside of your vehicle on a regular basis. If not, an air freshener can only attempt to cover up, not remedy, the cause of odors.

It's fine to use a particular air freshener scent. A multitude of such products are available at variety stores and auto parts houses. Solid types hang from knobs or are placed out of sight under seats. Liquids are also available in a variety of scents. Follow the directions on the label and don't use too much! Start out with just a hint of the product and gradually increase it to please your sense of smell.

Some car buffs clean the interiors of their rides with lemon-scented Pledge. The lemon scent stays with the car a long time, generating a pleasant aroma. They also dust vents using cotton swabs moistened with it. Others like to put a little Lysol in cleaning solutions and carpet shampoo to get similar results. These are the kinds of detailing tricks you can discover for yourself by visiting various car shows and rallies and talking with those folks who own automobiles detailed to your sense of perfection.

Chapter 4
Under The Hood
Preparations and Cautions

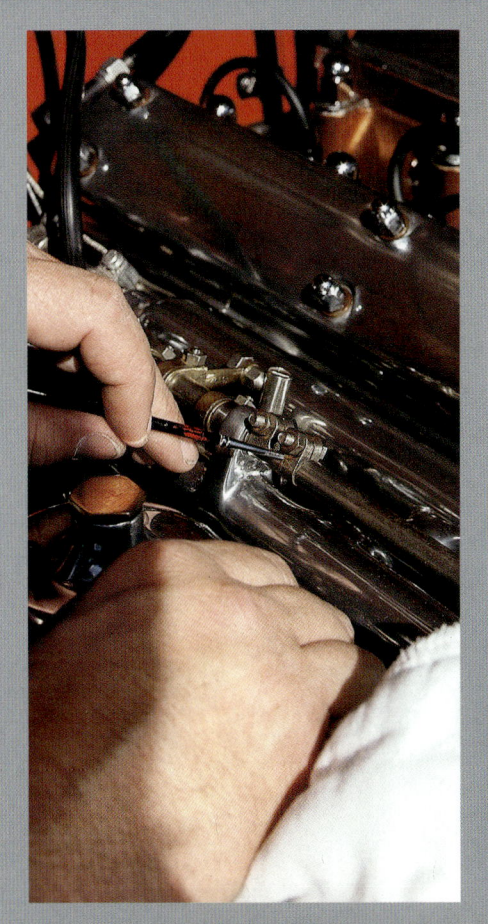

It's no myth. Clean engines are faster than dirty ones. They run cooler, too. And when someone asks to see under the hood, they'll be even more impressed with your car!

DETAILING Techniques

A clean engine compartment has benefits that extend beyond cosmetics. Repairs and maintenance are easier, problems are more readily detected, exposed linkages move more freely, and the engine might operate a bit cooler.

For full-scale detail projects on neglected vehicles, consider starting with engine compartment cleaning before others areas. Cleaning a dirty engine compartment is messy work, and chances are you will inadvertently splatter grease and grime on the fenders, cowling, and windshield.

The complete detailing of a neglected engine compartment that takes place just once a year or on a one-time basis may justify the use of high-pressure water spray and solvent-based cleaners. With this, you must be concerned about water and caustic cleaners being forced into electrical connections and ignition parts by the pressure washer. Controlled spraying and a planned approach should minimize problems.

Work Area

Engines that suffer years of grease and grime buildup should be cleaned at a place where residue is not a problem. A self-serve car wash with large drains and catch basins is ideal. Your driveway is not. Consider removing the big stuff at a self-serve car wash and performing the detailed cleaning at home. Detailed cleaning will not cause residue problems, because most grease and grime will be gone. You will focus attention on the smaller stuff, like wires, hoses, polishing, and towel cleaning.

The work site for detailed cleaning should provide a water source and plenty of light. Good water drainage should help keep your feet dry, and a source of electricity will allow use of a droplight for best visibility. For maximum brightness, line the inside of your droplight with aluminum foil. Place the shiny side out, and keep it away from the socket of the droplight to prevent a short.

Preparation

Consider initial cleaning the first step in detailing a neglected, grease-covered engine compartment. For this you will need a degreaser, brush, soap, old wash mitt, water source, and towels to cover and dry the distributor and carburetor. If you choose to work at a self-serve car wash, plan a time when you won't be hurried by a long line of other people waiting to use the stall.

Although covering the distributor is recommended, there are times when its exterior needs a bath in degreaser and water. In those cases, plan to dry it before attempting to restart the engine. A hair blow dryer works well, as do a clean dry towel and air pressure. To keep excess water from entering the distributor, remove the cap and place a piece of plastic over the entire exposed surface. Then snap the cap back in place. Water condensation may appear on the inside of the cap; dry it with a clean towel. After detailed cleaning, again remove all moisture in and on the distributor by drying with a towel, blow dryer, or air pressure.

Some detailers prefer to cover entire distributors and other electronic components and connectors with paper towels, folding and wadding them between ignition wires. The paper towels are then covered with aluminum foil, using tape

A nicely detailed engine compartment is a pleasure to look at, and regular maintenance is easier and more pleasurable without a blanket of grease to contend with. Wiping down the compartment frequently with a cloth and cleaner will help it stay clean and look good.

Under The Hood 63

Above Left: If your vehicle's engine compartment has been neglected and is covered with a lot of grease and grime, consider cleaning it initially at a self-serve car wash. Drains at most of these facilities are equipped to handle heavy grease and grime runoff.

Above Right: For those engine compartments in need of an aggressive initial cleaning, cover the carburetor and distributor cap after the air cleaner has been removed. Detailers have used rags, plastic, and aluminum foil to cover these parts. Covers are held in place with masking tape.

as needed. The paper towels quickly absorb moisture that seeps around the aluminum foil. You can also cover the distributor with a thick plastic bag. There must be enough slack in the ignition wires to allow the bag to fit over the entire distributor.

No water should be allowed to enter the carburetor. With the air cleaner off, gently fill the throat with rags and cover with a plastic bag taped to the top outer ring. If tape won't stick to the greasy surface, tie it with a string. Aluminum foil may be used in lieu of plastic; any method is good, as long as it keeps water out. This same philosophy holds true when removing air intake hoses and components for fuel-injected models.

Use a strip of duct tape to cover holes in the tops of battery caps. Duct tape also works well to cover holes in valve covers left open by the removal of air cleaner hoses. Other electronic parts may be protected as you see fit; use plastic wrap, paper towels, rags, tape, foil, or anything else you feel will keep parts dry.

A hair blow dryer works well to dry out the inside of distributor caps. Plan to do this after using a lot of soap and water to clean the engine compartment.

TECHNIQUE 19 INITIAL CLEANING

Time: 1/2 to 1-1/2 hours

Tools: Degreaser, cleaner, spray bottle, brushes, towels, safety goggles, rubber gloves

Talent: ★★↗

Tab: $10–$25

Tip: Place a large sheet of plastic under the vehicle to catch grease and grime residue

Gain: Much easier to work on the engine and its accessories

Complementary project: Wash the entire vehicle

Auto parts stores carry a variety of engine degreasers. Gunk in the spray can is easy to apply and removes grease handily; it is also available in quart and gallon containers. Solvent and kerosene work equally well. Many auto enthusiasts have had the best results buying solvent in the gallon can and applying it through a spray bottle. The nozzle on the bottle adjusts from a spray to a straight stream, and a clear bottle shows when solvent is running low.

Let the engine warm up to operating temperature before starting. Grease is easier to remove from a hot engine. With the air cleaner in place, spray the engine and all greasy spots with solvent. Let it soak into the hot grease for a couple of minutes. Then use water pressure to rinse clean. Aim the nozzle in such a way that water and grease don't splatter you. Because of the degreaser's powerful cleaning ability, consider wearing safety goggles or glasses to protect your eyes, and rubber gloves to prevent your hands from drying out. Remember that the engine and radiator will be hot, so watch where you rest your arms and hands.

After rinsing, look closely at the engine. Note remaining spots of grease and spray it again. Use a paintbrush or a parts cleaning brush on stubborn spots to work in the degreaser and help loosen buildup. Spray under the air cleaner and behind the power steering unit and alternator. Rinse with water. Next, spray the entire engine compartment with a cleaner like Simple Green or Meguiar's Extra. Use an old wash mitt on the air cleaner, fenderwells, firewall, and radiator. Rinse with water.

Two initial applications of degreaser and one of soap should have removed heavy accumulations of grease and grime. At this time, dry the top of the air cleaner and remove it from the engine. Have you covered the carburetor as earlier described? Concentrate on one side of the engine, spraying degreaser on all remaining rough spots. Use the paintbrush or parts brush freely. Look at areas around spark plugs, bell housing, and air conditioning units. Stick your head inside the engine compartment and look for grease. Spray, brush, and rinse with water until all grease is gone. Do this around the entire engine, including the intake manifold and carburetor.

To be assured of a grease-free engine, spray everything with Simple Green, Meguiar's Extra, or a cleaner of your choice. Use the paintbrush and wash mitt to remove dirt from around the battery, windshield washer bucket, compartment edges, and fan. Rinse with water. Continue until you are satisfied that the engine compartment is as clean as it is going to get.

Cautions

While it's easier to remove grease and dirt from an engine that has been warmed to its operating temperature, this procedure is *not* suited for all engines.

Cleaning an engine and its surrounding compartment requires a degreaser, an all-purpose cleaner, brushes, and rags or paper towels. You'll need masking tape and material to cover the carburetor and distributor. When cleaning is complete, you may want to touch up the paint.

Turbochargers and headers get extremely hot, even at normal operating temperatures. Spraying cold water on them while they are hot may cause cracks, warping, and other damage. Clean turbocharged engines and those with headers while they are cold. If you had to drive to a self-serve car wash, let the vehicle sit and don't start cleaning until the turbocharger or headers have cooled to the point that you can safely touch them with your hand.

Strong degreasers and powerful water spray will easily remove stickers and decals. To keep an engine looking original, factory-installed stickers should be kept in place. Gently use a paintbrush to clean them, and rinse with the nozzle at a distance to reduce the water's force.

High-pressure water can also cause other problems. Water, cleaner, and degreaser products can be forced into electrical connections to create a potential corrosion hazard. Improperly applied, high water pressure can also peel paint and cause other damage.

Engines can be cleaned using low water pressure—high-pressure water just helps to make the job go faster. For engines in high-performance cars, classics, and those in which high pressure may present special problems, use a garden hose. Lower the pressure by closing the spigot. As you rinse, agitate dirty areas with a paintbrush to help float away grease and grime.

Fenders and cowlings are protected with towels. Before you start cleaning, drape towels over those parts you want covered. A light water spray on towels helps keep them in place.

As with every other part of detailing, always start engine cleaning using the mildest methods. If your engine is not caked with grease, don't use a pressure washer or degreaser. Try a mild cleaner. Simple Green and Meguiar's Extra do not streak or cause corrosion. Mild dishwashing detergents have also been used with good results, while heavy-duty industrial strength cleaners have been known to fade paint and dry wiring. Use paper towels to remove the big stuff and rags for the rest.

Hood Underside

The underside of the hood should be the last engine compartment item to clean. The reason is simple. If you washed it first, drops of water would fall on your head all the while you were busy cleaning the engine. If the hood and engine both need lots of cleaning, do the engine first. Then, while the engine is still wet, wash the hood. Rinse the engine afterward to remove residue. If the hood is really dirty, consider covering the engine with a tarp or piece of plastic to prevent grease and grime from falling on the engine and into the engine compartment. Or you can certainly elect to wash the hood first and then dry it so that water drops don't fall on your head while cleaning around the rest of the engine compartment.

1. If you decide to employ high-pressure water to help you in cleaning an engine compartment, control the spray. Do not aim water directly at electrical components and be alert to water ricochet. It is best to wear safety goggles and rubber gloves for this project.

2. After the big stuff has been washed off, concentrate your efforts on the smaller things. Work to remove pockets of caked-on crud with a toothbrush or small parts brush. Spray solvent or an all-purpose cleaner on the spot, and agitate it with the brush to break it loose.

66 Detailing Techniques

During detailed cleaning, since there will be little water use, clean and dry the hood first and follow with the engine. This way, you won't have to contend with residue falling off the hood and onto the clean engine.

Drying

To dry an engine compartment quickly after an initial cleaning, start it up and let it idle. Check the water temperature gauge periodically to ensure the engine does not overheat.

Large pockets of water on the intake manifold are dried with a towel or wet/dry vacuum. The air cleaner should have been replaced, allowing it to dry. Water spots will be noticeable afterward but are easily cleaned with a damp towel during detailed cleaning.

Operating the engine will force water out of hidden pockets on the front of the engine. This will also give the alternator a chance to dry, as well as all the other engine components. The combination of forced air from the fan and normal operating heat should dry the engine compartment in 10 to 15 minutes.

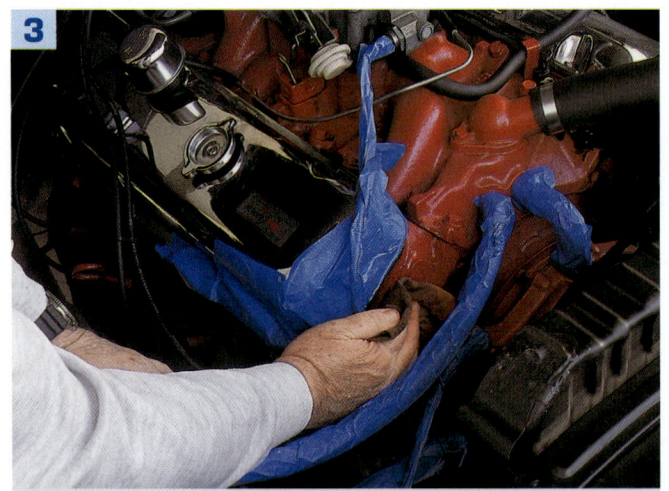

3. Start at the top of the engine and work down, taking time to inspect each part of the engine along the way. Reach into lower areas to remove grease and grime from all parts of the engine. As rags or paper towels become saturated, toss them out and use clean ones.

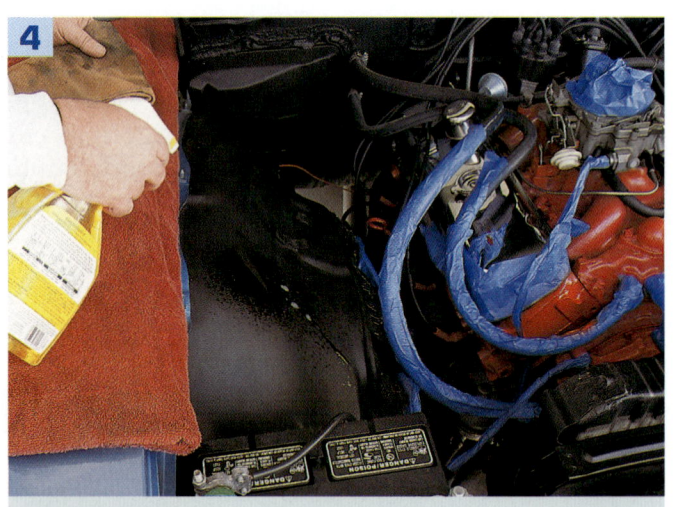

4. With the engine clean, focus your attention on the firewall and inner fenders. Note how this detailer has covered the fender with a large cloth to protect it from his body leaning against it and from solvents and cleaning residue.

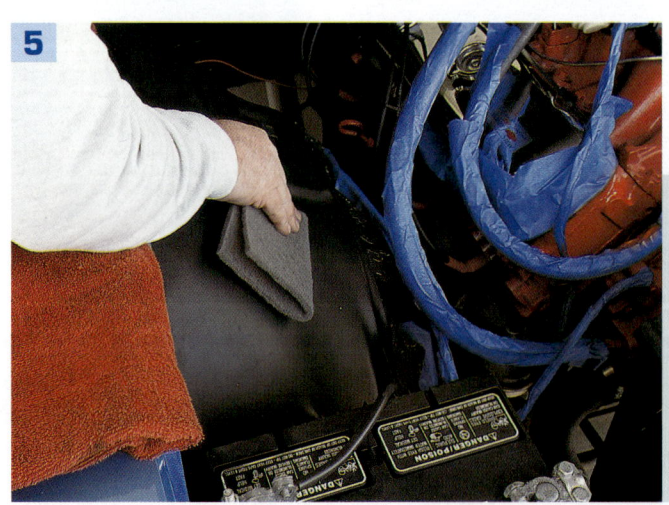

5. A very mild scouring pad may be needed to remove stubborn accumulations of grime buildup. Plenty of all-purpose cleaner must be sprayed onto the surface before using any type of an abrasive pad. Along with added cleaning strength, cleaner also serves as somewhat of a lubricant for the scouring pad to protect paint.

Under The Hood 67

TECHNIQUE 20 INITIAL CLEANING WITHOUT WATER

Time: 4 to 8-plus hours

Tools: Paper towels, rags, degreaser, cleaner, brushes, old toothbrush, small flat-blade screwdriver, dull putty knife, spray bottle, rubber gloves

Talent: ★★

Tab: $15–$25

Tip: Have a trashcan handy for the greasy paper towels and clumps of grease removed

Gain: Clean engine without fear of water damage

Complementary project: Remove, clean, and repaint engine compartment accessory parts

Occasions arise when owners of classic automobiles prefer to clean neglected engines without the use of water spray. Such may be when one purchases a neglected used classic and wants to make the engine compartment look new but fears that extensive water spray could harm delicate engine components or the originality of the factory stock engine underneath all of the grease and grime. Other times may simply stem from the fact that detailers don't want to go to a self-serve car wash or contend with the results of a driveway covered in grease. And there are those who prefer to do the work in a garage during an otherwise dull winter season.

Whatever the reasons, engines and engine compartments may be initially cleaned without the heavy use of water spray. This project will take time, patience, and a few rolls of paper towels.

Start with the hood. Spray a small section of the hood underside with degreaser, about 1 square foot. Use paper towels to wipe away the loosened residue. Go over that section again until it is clean down to the paint. Then move on to another square-foot section. Work your way around the hood until all grease and grime have been removed.

Once the grease has been lifted from the entire hood underside, spray a quarter of it with an all-purpose cleaner, like Simple Green. Use paper towels to remove residue once again. Repeat the process until the hood underside is completely clean. If the hood of your vehicle is covered with insulation, be careful not to disrupt it. Or, if it is in bad shape, plan to remove it all and replace it later, once the engine compartment has been completely detailed.

With the hood clean, focus your efforts on one part of the engine compartment—just the front, one side, the top of the engine, or the like. Set up a cleaning pattern—a place to start, the middle, and the end. In other words, put together a plan. Filthy engines offer all kinds of things to clean. Unless you have a master plan, you'll likely end up moving around all

To start a waterless engine compartment cleaning project, think about using compressed air to blow off all of the loose stuff. Do this with relatively low pressure and hold a vacuum cleaner hose close by to catch the debris. This will help to keep dust down and prevent leaves and other particles from flying all around your workspace.

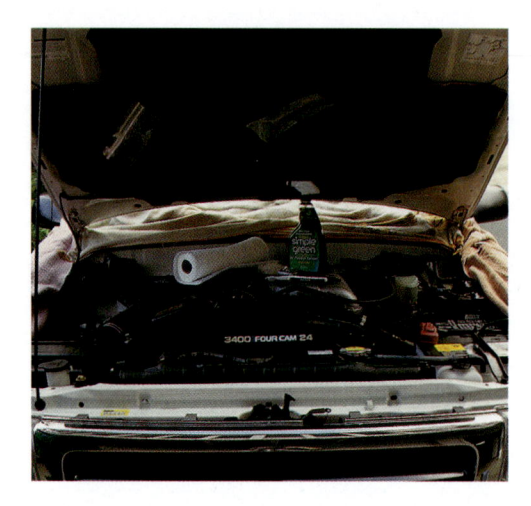

68 *Detailing Techniques*

over the place, cleaning here and there without making systematic progress.

It is best to focus on a spot, get it clean, and then move on to the next spot. This way you accomplish significant progress without having to repeat your efforts over and over again in the same general area. For example, clean the bulk of grease and grime off the air cleaner. Then remove it for detailed cleaning out of the vehicle. Once it is done, focus on the carburetor and its linkages, using degreaser, paper towels, and brushes as needed to remove all accumulations of grime. Residual degreaser that flows off the carburetor and onto the intake manifold will actually start to loosen grease there.

Move on to the intake manifold, then the heads, front of the engine, sides, and so on. Use a dull putty knife to gently scrape off large accumulations of grease from unpainted parts. A plastic putty knife may work well to remove accumulations from painted things, like the valve covers. Use an old toothbrush or parts cleaning brush to agitate degreaser on stubborn accumulations. A small flat-blade screwdriver works well to scrape away grease from tight spots found all over unpainted engine block castings. Pick up puddles of degreaser residue and globs of grease with paper towels.

Once you have removed the bulk of the grease, use the all-purpose cleaner to remove other major accumulations of dirt and grit-laden grime. Use paper towels extensively, as they are cheap and plentiful. Save rags for more intricate cleaning, when you need something that holds together better than the paper towels.

Once all of the parts on the engine are clean, look toward the battery area. Consider removing the battery for cleaning outside the vehicle. The space left open by its removal makes it easier to reach around the area for the easiest and most efficient cleaning. Do this around the entire compartment, taking your time to focus on the big stuff for now. After you have gotten the space clean enough to actually see what was lying underneath that blanket of grease, you'll be ready to tackle the next project of intricate cleaning.

Far Left: Spray a small section of the hood with degreaser or all-purpose cleaner. Use paper towels to wipe away the residue. Continue with small sections at a time to stay ahead of the project and minimize the amount of runoff falling onto the engine.

Left: Some parts around an engine compartment may simply be covered with road dirt. A dry paintbrush works well to dislodge this kind of buildup. It is easily whisked away with a damp paper towel or even a brush attachment on a vacuum cleaner.

As you come across pockets of caked-on crud, spray them with cleaner and agitate with a brush. Use a paper towel or rag to wipe off the residue. Take your time and stay focused on one general spot until it is completely clean before moving on.

TECHNIQUE 21 DETAILED CLEANING

Time: 1 to 4-plus hours

Tools: Cleaner, old wash mitt, paper towels, rags, spray bottle, brushes, paintbrush, scouring pads, small wire brush, degreaser, rubber gloves

Talent: ★★★

Tab: $10–$25

Tip: Ensure you have good lighting under the hood

Gain: Much nicer to perform routine maintenance on a clean engine

Complementary project: Paint the engine block and other compartment accessories

Starting at the hood, don't forget to clean the hood hinges and supports. Apply cleaner or solvent on a rag or paper towel away from the vehicle to prevent drips from falling onto engine compartment parts. Use a toothbrush or parts brush to break loose tough buildup.

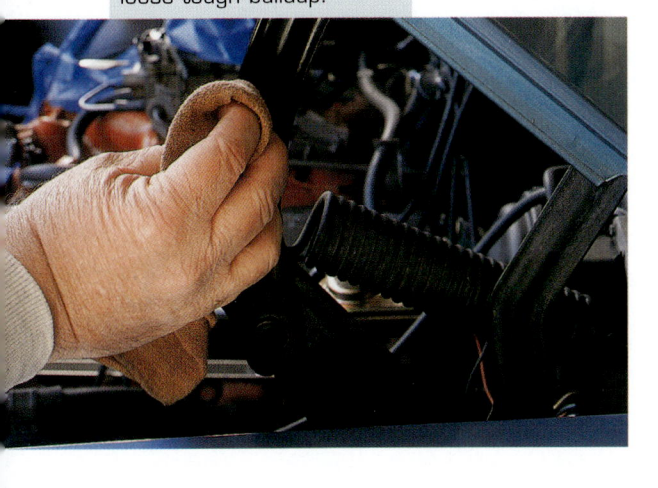

Engine

Start out by cleaning and drying the underside of the hood first. Spray about a quarter of it with cleaner and use paper towels to wipe off the residue. Then spray cleaner on a soft cloth and go over the area again to pick up missed spots. Inspect that section closely to ensure you removed dirt from crevices and valleys created by the design of the hood structure. Use caution around those openings commonly found on inner hood supports. Their edges can be very sharp.

After the hood is done, remove the air cleaner. To help reach every part of the engine compartment, remove the windshield washer fluid reservoir, radiator overfill container, and battery. When disconnecting the battery, take off the negative lead first, positive last. These parts will be cleaned outside of the vehicle using cleaner and water. If need be, scrub with a soft-bristled brush.

Clean the carburetor first. Should cleaner not do the job sufficiently, spray stubborn spots with degreaser and agitate grimy accumulations with a brush. Remove residue with paper towels or a rag. Use the toothbrush around linkages and mounting screws. Concentrate your efforts on one spot at a time; don't get ahead of yourself. When the carburetor is clean, move to the intake manifold. SOS pads make quick work of removing caked-on grime. Use just enough pressure to get the dirt off and not scar paint.

Next are the head and spark plugs. Remove only one spark plug wire at a time to guarantee replacement in the proper sequence. Clean as needed, using the toothbrush, SOS pads, paintbrush, and cleaner. Spark plug areas in the head are cleaned with the cable-cleaning end of a battery brush or small wire brush. Greasy wires are cleaned by dabbing a bit of solvent on a rag and wiping. Follow with a soapy rag to remove solvent residue.

At the front of the engine, use an old wash mitt dipped and wrung out with a cleaner solution in a bucket to clean fan blades and housing parts. Take pains to remove all signs of dirt around the thermostat housing and water pump. Use SOS pads on unpainted alternator and support brackets. Most bare metal parts, such as fuel lines and carburetors, will clean nicely with scouring pads. They may be polished later with a chrome polish of choice.

Chrome headers should be washed with soap and water, and then they should be polished with Happich Simichrome or another comparable chrome polish. You can wax for protection, even though wax won't last long under the high heat that headers produce. Stainless headers may be polished, too. Painted headers are cleaned with soap and water. Touch-up is done with the appropriate color of heat-resistant paint, using paint blocks, towels, and masking as needed.

Cleaning an engine that is not painted may require more work because you can't rely on paint to cover up tiny flaws. Use a toothbrush, small wire brush, and scouring pad to reach into grooves and pockets formed by the engine design. Do not use harsh scouring pads on polished aluminum heads. For those, try a lot of patience and a toothbrush with Simple Green, Meguiar's Extra, or a comparable cleaner. Stains are polished out with products designed for aluminum polishing. A few detailers have found good results using mag cleaner on aluminum engine parts. Although the cleaner quickly removes grease and stains, you must be concerned that it may cause corrosion on other parts, such as wires, hoses, and electrical components. During such cleaning with harsh chemicals, protect these parts.

Firewall Area

Most vehicle firewalls are covered with all sorts of attached parts and accessories. You'll have to exercise patience in cleaning this area. Spray a section with cleaner and put your hand in an old wash mitt for your first cleaning effort. This should result in the major dirt accumulations being whisked away. Follow that with cleaner on a soft cloth. Fold, shove, and maneuver this cloth around and over obstacles to remove streaks of dirt and the remnants of grime deposits.

A small floppy paintbrush works well to agitate cleaner around the parts found attached to firewalls. Remove the residual wet accumulations with a paper towel or cloth. Stay at it until the section you are working on is clean. Go over areas more than once to ensure excellent results.

Parts that are easily removed should be taken out of the engine compartment for cleaning. It is much easier and much more efficient to clean plastic containers in a slop sink or wash bucket filled with sudsy cleaning solution. This will also allow you to clean the containers' support brackets and surrounding firewall area, too.

To clean windshield washer hoses and other hoses and wires on or near the firewall, spray cleaner on a small rag and then drape the rag over your hand. Grasp the hose or wire with the rag and pull your hand along it. This lets you clean all around hoses with just one maneuver. Go over hoses and wires a couple of times to ensure complete coverage. Move the rag around on your hand so that a clean section comes in contact with the hose or wire each time.

Move along the firewall to clean everything in sight. From the passenger side, use a floppy paintbrush to agitate cleaner around air-conditioning and heater components. Be sure to fold cleaning cloths as a means to reach inside the grooves and valleys on cowlings and on other firewall features. Move the droplight around to maximize your visibility along the lower parts of the firewall. Reach as deep as you can for optimum cleaning.

Fenderwells

The same general process employed to clean the firewall should be used for the fenderwells. Some vehicles feature wide-open fenderwell areas to make cleaning a snap. Others are chock full of parts to make cleaning a real chore. Take your time and be patient.

Start at the front or back of a fenderwell area and concentrate on just a small section at a time. Basically touch every square inch of the fenderwell and every part attached to it. Use a paintbrush to agitate cleaner around parts and bracket supports. Don't forget to clean under hoses and wires.

If parts attached to fenderwells are easy to loosen, go ahead and do it. This will allow you to clean them more

Above Left: Remove the battery and clean it outside of the vehicle. Use a mixture of baking soda and water to dissolve corrosion; a few teaspoons of baking soda stirred into a glass of water will work fine. Be sure to brighten up the battery posts, too. Take a look at the battery tray once the battery is removed. If its condition is poor, replace it with a new one; they generally cost around $15 to $20. If it is in good shape, clean it and plan to give it a new coat of paint.

Above Right: Start engine cleaning from the top and work down. Concentrate on the carburetor and the linkages connected to it. Spray degreaser or all-purpose cleaner onto the parts that need cleaning; agitate the mix with a paintbrush or mild scrub brush and remove residue with a paper towel or rag.

Under The Hood

Once the engine has been cleaned completely, focus attention on other surrounding parts. Spray a dirty part with solvent or cleaner and let the mixture sit for a few moments to soak in. The big stuff that loosens up right away is removed with a rag or paper towel. When the initial spray-and-wipe process fails to remove all dirt and grime, spray again and agitate the mixture with a paintbrush or parts brush. This works well to break loose accumulations of dirt and grime from corners, recesses, grooves, slots, and the like.

Below Left: After agitating solvent or cleaner on a part, use a rag or paper towel to wipe away the residue. Use your finger or the handle of a toothbrush inside the rag to reach into recesses to pick up cleaner and dirt runoff.

Below Right: Hoses and wires need cleaning, too. Although this hose appeared clean to start with, you can see how much grime came off it with just a couple of swipes of a cleaner-dampened cloth. Hoses and wires must be clean before they will accept a light coat of dressing to look new.

completely and also wipe off the fenderwell body underneath. Spray cleaner directly onto stubborn dirt accumulations, and use a soft toothbrush to break that stuff loose. As you notice dirty spots where you have already cleaned near the area you are working on, take the time to clean them. Don't rely on your memory to come back later and get them.

Grille and Radiator Area

At the front of the engine compartment, you are likely to find a host of small areas stained with dirt deposits. Such are the corners of radiator brackets, seams along radiator support brackets, behind the headlights, and so on. There shouldn't be much in the way of grease deposits, but dirt accumulations are common.

Since there are so many sharp edges in this area, plan to initially clean the space with a wash mitt dipped in a bucket of soap and water and wrung out. You shouldn't need to have the mitt saturated and dripping. Put your hand into the mitt and work around a specific area. Follow up with a cloth lightly sprayed with cleaner. Use a small paintbrush to loosen dirt accumulations in corners and along valleys. Follow up with a small towel dampened with clear water to wipe away cleaner and remaining dirt residue. Don't forget to wipe down the fan blades, too.

Once the areas in front of and around the radiator are clean, you might take the time to inspect the radiator fins. Use needle-nose pliers or the end of a paper clip to pull out the remnants of bugs stuck in the fins. The Eastwood Company even offers special pliers designed to assist you in straightening out any bent radiator fins.

Battery Box

A mixture of baking soda and water works well to remove the buildup of battery acid residue, which is the foamy looking white stuff that accumulates around battery terminals and hold-down brackets. A few spoonfuls of baking soda in a small glass of water should work fine. Apply the solution freely, working it into corners with a paintbrush or old toothbrush. Rinse thoroughly with water. Continue the process until the mixture no longer fizzes and bubbles. Do not use any of the rags from battery cleaning on any other part of the vehicle, as acid residue on the rags will cause unwanted damage. Just throw them away.

The area below the battery box also needs attention, as it likely suffers from the same type of corrosion. Clean threaded support rods with the same solution. Clean them with a toothbrush and scouring pad as needed. If the entire box is easily removed, take it out. You can clean and paint it as its condition warrants. If you plan to paint the battery box, it is best to use a paint product especially made for battery box/tray painting. Battery box paint is available through The Eastwood Company and most auto parts and auto-body paint and supply stores.

When the detailed cleaning is complete and the paint dry, line the bottom of the battery box with a ribbed rubber mat. Be sure to make drain holes in the mat so that water will not puddle under the battery.

TECHNIQUE 22 PAINTING THE ENGINE BLOCK AND OTHER ENGINE PARTS

Time: 1 to 4 hours

Tools: Engine paint, paint block, rags, masking paper and tape, lacquer thinner, paint thinner

Talent: ★★↘

Tab: $20

Tip: Wear lightweight Latex gloves to prevent paint-stained fingers

Gain: New-looking engine

Complementary project: Replace old worn stickers and decals with new ones

Auto enthusiasts have mixed feelings about painting engines. Some prefer engines that look original and don't care if a little paint is faded or chipped. They consider those kinds of flaws on a clean engine authentic and insist they give the impression of an engine that has been cared for over the long run. Others prefer blocks, heads, and manifolds that sparkle.

Most agree, however, that an unusually high-gloss detail might tend to conjure thoughts of engine neglect that was glossed over with a quick cosmetic fix. A happy medium may be thorough cleaning with judicious painting. Once again, you must make those determinations based on the purpose of the vehicle you are detailing.

Engine paint is available at auto parts stores. It comes in various colors, such as Chevrolet Orange and Ford Blue. The paint is specially formulated to withstand the high temperatures experienced with engine operations. Buy two cans of paint if you intend to paint the entire engine. Warm cans of spray paint in a sink of hot water for about five minutes before starting. This will help to thin the paint to make for better mixing and smoother application.

Warnings on cans of spray paint advise not to subject them and their contents to temperatures over 120 degrees Fahrenheit. Hot water used to warm cans should be from the tap and not hot enough to hurt your hand. Spray paint works much better at 80 to 90 degrees than at 50 or 60 degrees Fahrenheit.

The engine, with air cleaner attached, should be allowed to idle up to operating temperature before you begin painting. A warm engine allows paint to spread quickly and evenly, reducing runs and making for quick drying.

Take the air cleaner off and paint it away from the car. Gloss or semigloss black looks good. The wing nuts can be painted bright silver. Paint them while they are still warm. Air cleaners not needing paint and those of unique colors may be waxed after the engine has been completed.

Before spraying paint on the engine, aim the can away from the vehicle and test the spray pattern. Assured that the spray pattern is good, start painting the engine at the intake manifold. Disconnect throttle return springs and move the wires out of the way. When that spot is painted, replace the wires and springs and move on to the next section. By doing a little at a time, disassembly will be minimal and reassembly should not be complicated. The object is to make the engine look its best without having to dismantle more than necessary.

Don't be overly concerned with small patches of overspray, as they will be removed later with lacquer thinner and a rag. At this stage, be concerned about complete paint coverage on all parts of the intake manifold. Use a paint block (a piece of thin cardboard the size of a license plate) to protect parts, such as the carburetor, alternator, and hose connections. Use masking tape to protect unpainted valve cover gaskets, fuel lines, and anything else desired.

Masking tape will adhere to clean engine parts and accessories. Blue masking tape is designed to come off much easier than the standard tan-colored masking tape. Mask off everything on and around the engine that you do not want painted and that cannot be held out of the way while spraying paint.

Under The Hood 73

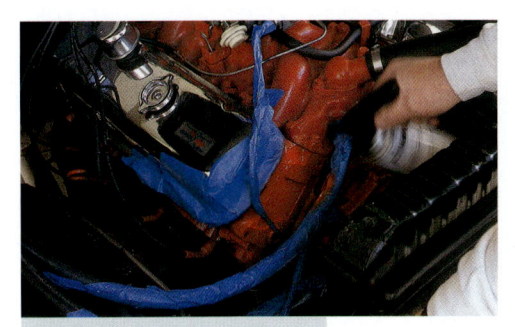

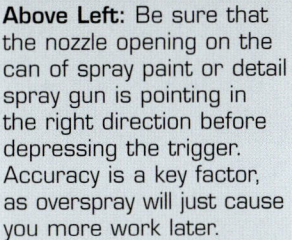

Above Left: Be sure that the nozzle opening on the can of spray paint or detail spray gun is pointing in the right direction before depressing the trigger. Accuracy is a key factor, as overspray will just cause you more work later.

Above Right: To get paint into the tight recesses around engine blocks and manifolds, some detailers have enjoyed good results placing a small plastic tube inside the spray can's nozzle opening. These tubes are most common with cans of WD-40. The concentrated paint spray will go right where you want it.

Test a quick spray first outside of the engine compartment to ensure the spray pattern is correct. Then point the nozzle at the correct spot and spray paint in short bursts. It is much better to apply two or three light coats than a single heavy one that will likely result in paint runs.

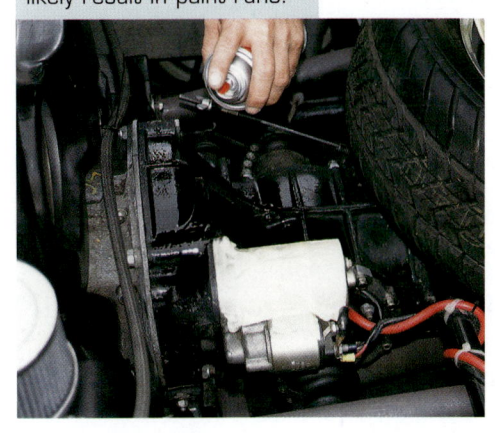

When the manifold is done, move to the valve covers and down the heads. Hold wires out of the way. Since the hot engine will cause paint to dry quickly, lay the wires down after a minute or so. Maintain smooth, even strokes with each pass of the paint can. Each coat should be thinly applied, allowed to dry, and retouched as necessary. As you paint the head, pull plug wires one at a time, replacing before pulling another. The spark plugs can be removed, or you can cover them with a short section of garden hose.

Continue painting around the engine until it looks the way you want. The front of the block will pose some problems. Turn the paint can upside down to reach lower areas. If the can is running low on paint and won't spray in that position, use a new can. Use a paint block around the fan belt, power steering unit, radiator hose, and anywhere else you need to.

Engine Compartment Parts

The best way to paint engine parts is to remove them from the engine, sand as needed, prime, and paint. If you are inclined to make the engine look its best with the least amount of work, you can paint parts while they are attached to the engine.

Such parts are the alternator, power steering and air-conditioning brackets, brake fluid reservoirs, battery boxes, and fan blades. Use gloss or semigloss black paint or original colored paint purchased from The Eastwood Company or an autobody paint and supply store. A paint block and towels work well to protect the engine, hoses, and wiring from overspray.

The same paint can be used to rejuvenate exposed frame members, motor mounts, the steering box, and the radiator. Cleanup is the same as for engine painting—lacquer thinner on metal parts and paint thinner on stickers or decals.

Using a droplight, carefully scan the engine compartment, looking for parts in need of paint touchup. As you come across them, use masking tape as necessary to protect adjacent items from overspray. You can also lay rags or towels over items for paint overspray protection. With a light touch, depress the paint can or paint gun nozzle control and apply paint. Keep the nozzle moving, as holding it steady in one spot will quickly produce paint runs. Opt for two or three light passes rather than a single heavy application.

In lieu of masking, you can strategically place a thin paint block (a piece of thin cardboard the size of a license plate) between the part to be painted and nearby obstacles, such as an alternator bracket. Put the edge of the paint block next to the bracket and ensure it covers the part you do not want painted. Then judiciously spray paint with just quick spurts. Other parts easily painted with the use of a paint block are radiators, black fan blades, front-end suspension members, and the like.

Small parts taken off of the engine also may be painted. Paint the wing nut on the air cleaner and older radiator caps bright silver. Use gold paint from The Eastwood Company for brake fluid reservoirs. Shoot black air cleaners with a semigloss black paint to make them look new.

Engine Compartment Firewall and Inner Fenders

Do not attempt to paint the firewall and fenders unless they are in desperate need of paint and are black to begin with. The only way to professionally paint an engine compartment firewall and fenders a color other than black is to remove the engine

Detailing Techniques

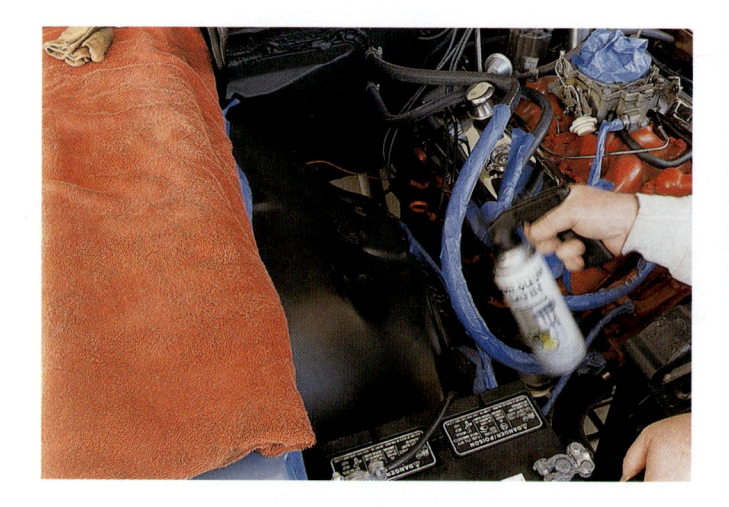

and start from scratch. Black is forgiving and won't emphasize minor flaws. Even at that, consider saving this type of paint job for a regular driver, as opposed to a more special automobile. Or consider painting as a temporary measure on vehicles slated for future restoration.

Should you choose to paint the firewall and fenders on your vehicle black, start out by masking the top edges of the fenders and cowling. Mask the voltage regulator and other parts attached to the firewall and inner fenders, too. Cover the engine with masking paper or towels. You can use newspaper to mask, but make sure it is at least two sheets thick, as paint will bleed through single newspaper pages. Remove wiring from holders along the fenders.

Start painting at the firewall and work your way around the compartment. Hold wiring out of the way with your free hand and protect other parts with a paint block. Apply paint with smooth, even strokes, holding the can at least 6 to 8 inches away from the surface. Overspray is removed with lacquer thinner, except on stickers and decals. Use paint thinner on those and employ a very soft touch to prevent rubbing so hard that the printing on the sticker or decal is faded.

As with the firewall and fenders, paint the underside of the hood black. Mask off louver and scoop openings. Place towels, masking paper, or double-thick newspaper over the engine, fenders, cowling, and windshield. Apply paint according to instructions, with smooth, steady passes. Opt for two or three light coats, as opposed to a single heavy application. This will go a long way toward reducing problems associated with paint runs. Plan to overlap each pass by 1 or 2 inches.

Removing Paint Overspray

Paint overspray is easily removed with a rag dampened with lacquer thinner. Stick your finger into a rag and moisten it with a dab of lacquer thinner. Do this away from the vehicle, as lacquer thinner will almost immediately blemish paint it spills or drips onto.

Gently rub spots of paint overspray until they vanish. If carburetor springs are coated, take them off of the linkages and clean them out of and away from the engine compartment. Do the same with other parts that are easily removed.

If paint somehow ended up on paper or vinyl decals or stickers, *do not* use lacquer thinner for overspray removal. Lacquer thinner is much too strong and will quickly remove any printing that was on the sticker or decal. Instead, try using a very light touch with a rag dampened with paint thinner. Start exceptionally light and apply more pressure to the overspray as needed. Keep a sharp eye out for any problems that creep up with regard to the sticker or decal printing starting to fade.

Take your time and thoroughly inspect parts around the engine block to ensure all overspray is removed. Use a droplight to aid this inspection.

Above Left: Painting firewalls and inner fenders is easiest when those areas are black. Black is very forgiving. Be certain to cover the tops of fenders and the cowling to guard against overspray on those surfaces.

Above Right: As with engine block painting, you should take time to mask off spots on engine compartment parts that you do not want painted. Good masking protects against overspray and also makes it easier for you to apply paint.

Specialty paint for engine compartment parts is available through autobody paint and supply stores and through The Eastwood Company. Along with battery tray coating, specific colors and paint mixes are offered for headers, chassis, radiators, alternators, and so on. The colors, textures, and formulas are designed to make parts look new and hold up under the conditions they are exposed to.

TECHNIQUE 23 POLISH AND SHINE

Time: 1/2 to 1 hour

Tools: Metal polish, paint polish, wax, vinyl/rubber dressing, soft cloths

Talent: ★★

Tab: $5–$10

Tip: Place towels under parts to catch dry polish residue

Gain: Makes metal parts, hoses, and wires look new

Complementary project: Align and straighten hoses and wires and ensure they are properly placed in their correct bracket holders

A great deal of time was spent making this engine and its surroundings sparkle like this. As dried polish is wiped off of shiny parts, you may have to contend with powdery residue landing on other parts. Whisk it away with a dry paintbrush. Or you might try using polishing pads that leave behind just a haze, rather than dried paste powders.

Chrome accessories inside the engine compartment are polished and waxed just like other chrome pieces. Use Happich Simichrome or any other brand-name chrome or all-metal polish. Place a small dab of polish on a clean soft cloth and start rubbing. Wipe off dry polish with a clean side of the cloth. Stubborn spots of rust are removed with number 0000 steel wool and polish. Follow polishing with a light coat of wax for lasting protection.

You might consider using polishing pads in lieu of regular metal polish. Never Dull and Mothers Polish in a Pad are basically wads of soft clothlike material impregnated with liquid polish. Simply tear off a chunk and wipe it on the part needing polish. These products dry to a light haze without much, if any, dry powdery residue.

Hoses and wires should look new after a light application of vinyl/rubber dressing. Clean first as needed with a damp towel or cloth lightly sprayed with an all-purpose cleaner. Dry with a clean cloth and then spray dressing on a separate clean cloth and wipe on. Remove excess by wiping again with the dry side of the cloth. Do this on vacuum, radiator, and windshield washer hoses, as well as on ignition and other wires, and all other rubber parts in the engine compartment. The easiest way to accomplish this is to drape the dressing cloth on your hand and lightly dampen it with dressing. Then grasp the hose or wire and run your hand along it for complete coverage. Be sure to wipe off excess dressing before moving on to another part.

If the firewall and inner fenders display good paint, consider polishing and waxing. Use regular paint polish like Meguiar's Number 7 and regular carnauba-based car wax, or try a one-step product like Meguiar's Car Cleaner Wax. One-step products generally result in less powdery residue left over once the wax dries, and they also help to clean pesky spots of remaining dirt that may have been missed earlier. Wipe off any residual dry polish or wax powder with a clean cloth lightly dampened with water.

Except for the areas around turbochargers and headers, most engine compartments do not get hot enough to immediately destroy wax protection on painted surfaces. You can expect a wax job to usually last a month or so, depending on weather and vehicle use. Let your eyes be the judge. If the air cleaner, inner fenders, and firewall look dull, polish and wax. The same holds true for wires and hoses that might need treating with dressing.

Detailing Techniques

TECHNIQUE 24 EXTRAS

Time: 1-plus hours

Tools: Cleaner, polish, cloths

Talent: ★

Tab: $5–plus

Tip: Compare your vehicle with those seen at quality car shows

Gain: Maximizing the longevity of an engine compartment detail

Complementary project: Replace old-looking ignition wires

Generic auto detailers often spray entire engine compartments with crystal clear lacquer paint after everything else is done. This paint will make minor dirt stains disappear, hoses glisten, and wires shine. The whole compartment looks like new. But that look does not last long because lacquer yellows and causes hoses and wires to become brittle.

True auto enthusiasts shudder at the thought of coating an engine compartment with clear lacquer. They would never do such a thing. Clear lacquer is just a quick fix designed to make ordinary used cars look their best until buyers drive them off the dealer's lot. Unless you are just trying to spiff up a regular driver for a quick sale, shy away from using clear lacquer in the engine compartment. Exceptions may be hard-to-reach areas on regular drivers, such as lower engine compartment areas and the narrow but open space between the grille and the radiator.

One way to maintain an engine compartment detail is to clean the area each time you wash your vehicle. Use an older wash mitt dampened with soap and water from the wash bucket. Wipe off dust and then dry surfaces with a clean towel to remove streaks. Use paper towels and cleaner to whisk away the beginning smudges of grease and grime.

Focus your attention on the edges of the hood and around the fenders, where water may have splashed up from front tires running through road puddles. Wipe off hoses and wires by laying a damp cloth across your hand and then grabbing the hoses and wires and run-

Far Left: This engine compartment is clean and really shiny—maybe a little too shiny. Some detailers spray entire engine compartments with crystal clear lacquer to hide flaws and make everything look wet and glossy. This should not be done to special vehicles, classics, and those in good condition.

Left: The plastic and rubber parts in this engine compartment look good. They are clean. A light coat of vinyl/rubber protectant will make them shinier, but they look fine just being as clean as they can be.

Under The Hood

The best way to keep an engine compartment looking good is to clean it regularly. Just spray a small amount of cleaner onto a cloth and wipe away the beginning signs of dirt, grime, and grease buildup. If you were to take a few minutes to wipe down the easily accessible parts of the engine compartment every time you washed your car, it would stay looking nice for a long time.

ning your hand along their length. Dry with a clean cloth.

Once the compartment is clean, look closely for any painted surfaces that need polish and wax, rubber parts that need dressing, or chrome that needs polish. Be sure to check the underside of the hood, too.

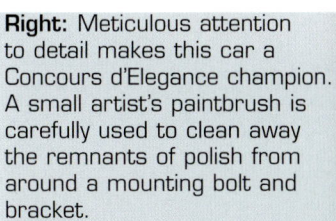

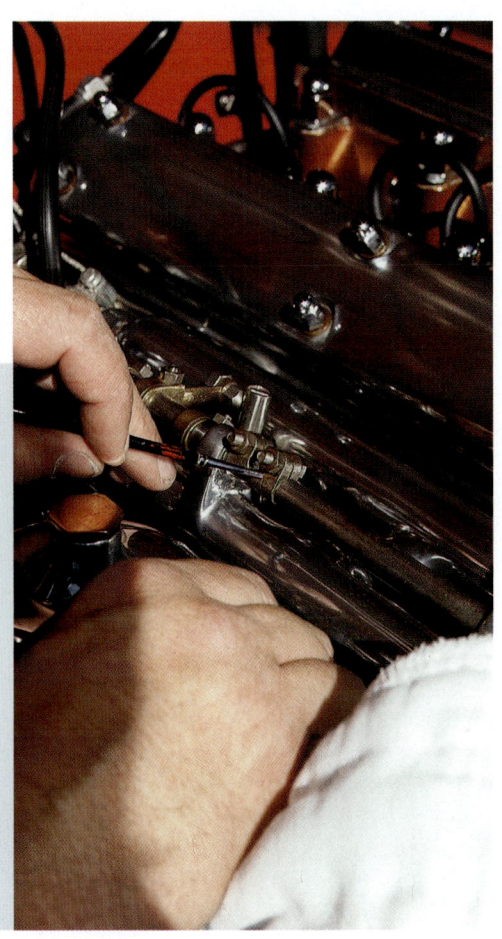

Right: Meticulous attention to detail makes this car a Concours d'Elegance champion. A small artist's paintbrush is carefully used to clean away the remnants of polish from around a mounting bolt and bracket.

Bottom Right: This concours-winning engine compartment stands tall and is definitely crisp. You can tell that the car is driven by the detection of bugs caught in the radiator fins. In fact, this car is frequently driven in slaloms at specific concours events. After the competition, you can bet that the radiator fins will be cleaned to perfection.

Decals and Emblems

Finishing touches to an engine compartment detail include lots of little things—straightening wires and hoses, aligning wires on support brackets, aligning hose clamps, polishing and waxing painted surfaces, and dressing plastic parts to make them look new. Plastic zip ties are excellent items to use for tying together bundles of wires and other such accessories to hold and keep them in an aligned manner. The more you tinker to straighten things out, the better the engine compartment on your automobile will look.

Decals and emblems are important pieces of equipment for classic cars and trucks. Most of the older models had engine compartments that featured various decals and emblems that denoted engine size and other engine traits. It is important that these factory stickers be replaced on these vehicles to help them maintain originality. Look on the Internet and through various automotive magazines to learn where you can purchase original equipment decals and stickers for your vehicle.

Chapter 5

EXTERIOR SHINE
Preliminary Basics—Polish vs. Wax

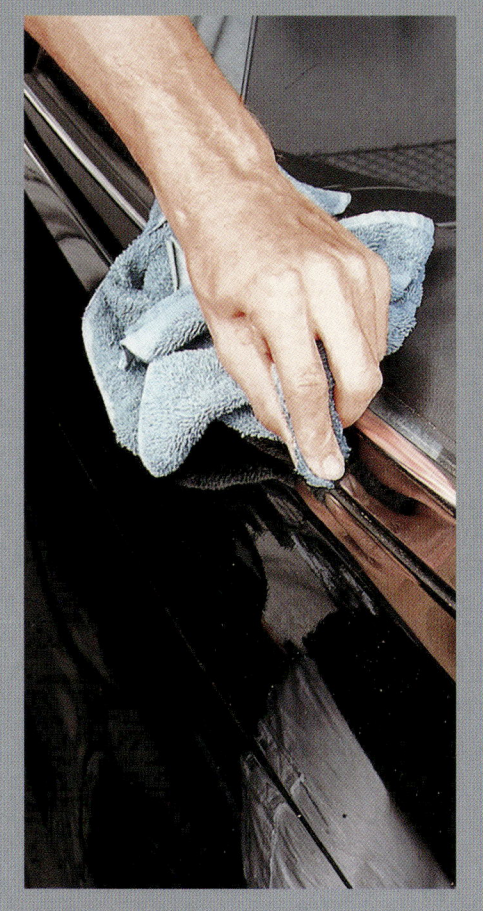

Image is everything. Here's how to polish and wax like a pro.

Detailing Techniques

Polish and wax are not the same. Polishes clean, and waxes protect. You cannot remove oxidation by using wax alone; it will smear and be almost impossible to remove from a severely oxidized painted surface. By the same token, once you remove the oxidation by polishing a painted surface to perfection, you cannot expect the shiny surface to remain that way for long unless it is protected with a coat of good carnauba-based wax. Just remember—polish first to clean and then wax to protect.

Polish

There are two types of polish: one for metal and one for paint. Chrome and all-metal polishes work to remove rust, pits, and blemishes. They contain very fine grit. Happich Simichrome and other name-brand mag and chrome polishes work well. Follow polishing with a light coat of carnauba-based wax for lasting protection.

Some paint polishes also contain grit, different from the grit in chrome polish. Other paint polishes rely on chemicals to polish and restore paint luster. Grit-containing polish comes in a variety of textures. Rubbing compound is by far the harshest and most potent; you should never have to apply rubbing compound except to the very roughest of surfaces. Polishing compound is similar but softer. It is needed on paint that is heavily oxidized, stained, and almost dead. Almost as a last resort, detailers will use polishing compound before recommending new paint jobs. Be careful using heavy polishing compounds, because you don't know how much paint is left on your vehicle. You can quickly polish right through the last layer of paint down to the primer, creating a major paint flaw that can be fixed only by a paint job.

Meguiar's, Mothers, Eagle One, and other manufacturers offer large assortments of automotive paint polishes. Most are available through auto parts houses, with the greatest variety found at autobody paint and supply stores. Determining the right product depends on the condition of the paint on your vehicle and whether you plan to use a machine buffer or employ hand application. The safest way to choose the right product is to ask a store clerk at an autobody paint and supply store or a professional detailer. If you want to choose the product yourself, read labels on the different products and opt for the one or two that contain the least grit for the type of oxidation problems you face. It is better to apply a mild polish two or three times than to use a polish that is too harsh a single time.

Wax

Choosing the right wax can be quite a dilemma. By far, the most common question is, "What is the best wax to use on my car?" The answer is not simple, nor is there common agreement among auto enthusiasts.

Some prefer the ease of application and minimal (if any) powdery residue left behind by one-step cleaner waxes. The drawbacks may be that they contain ingredients that can cause repaint repair jobs to fisheye and they do not last as long as carnauba-based products. Along with providing wax protection, they will remove small spots of road tar, bug splatter, and lots of other imperfections. Some car buffs wax a different portion of their cars with a one-step product each time after washing. They figure that each part of the vehicle will get lightly

Below Left: This is a three-phase polish and wax system from Mothers. First is the Pre-Wax Cleaner (polish) designed to remove oxidation and clean the paint's surface. Next is the Sealer and Glaze (polish) designed to add depth and shine. And last is the Pure Carnauba Wax designed to protect the beautiful shine developed by use of the previous two polishes.

Below Middle: Different types of polish are manufactured for bare metal as opposed to painted surfaces. This is an array of metal polishes designed for chrome, wheels, and all metals. All of them do a good job when their label instructions are followed.

Below Right: Most polishing product labels will list the types of paint problems they are designed to solve. This one is specifically made for machine (buffer) use. Not all polishes can be safely applied with buffers, and you must read labels to ensure compatibility with machine use.

Exterior Shine 81

Above Left: Once an automobile exterior has been polished to perfection, the detailer must protect the paint finish with a coat of wax. Without wax protection, paint finishes will quickly dull, fade, and oxidize. A good wax job should last from three to six months, depending on weather and exposure to the sun.

Above Middle: The right half of this hood has been polished and waxed, while the other has not. A good rule of thumb to remember is that when water stops beading tightly on the surface of the paint finish, it is time for a wax job.

Above Right: A good-looking paint job will remain looking good with proper care. Normally, a paint finish maintained in good condition with a wax job every three months or so will only require polishing about once a year. Polish will remove old wax and swirls to make the paint look new, and a fresh coat of wax will help keep it looking that way.

polished and waxed once a month, or so. Thus, they get the satisfaction of knowing that their ride is always clean and waxed.

Others prefer carnauba-based auto wax. They feel that the lasting quality is well worth the effort of application. Once paint is polished and in good condition, all it needs is protection. Once a year, these enthusiasts will polish their vehicles with a mild polish, like Meguiar's Number 7, to get rid of old wax and surface blemishes, and then wax with a carnauba-based product. Unless you have allowed the wax job on your car to fade by not waxing once every three months or so, and if the paint surface has not been subjected to any harsh pollutants, you should not have to polish more than once a year.

Auto parts stores are filled with racks of auto wax. Choosing the right one is difficult. You can try each brand until you find the one best suited for your needs. Talk to other car enthusiasts to find out what they like. There is no clear-cut definition of a perfect car wax. Most auto buffs have found their preferred brand by talking with other enthusiasts and trying different products until they find the one that seems to be the easiest to apply and remove and does the best job of making the paint on their car look its best.

Oxidation

When you wash your car, does paint rub off onto the wash mitt and drying towel? If so, the paint can be considered severely oxidized. If not, the paint may still suffer from moderate or light oxidation. You need to determine the degree of oxidation and work toward restoring the shine.

Oxidation starts on the top layer of paint, which dries out and becomes chalky. Paint should never really dry completely. Wax protects it from the sun's ultraviolet rays and the elements and also helps it to retain certain oils that paint needs to stay healthy. With little or no protection, layers eventually dry out, lose oils, and therefore lose the ability to shine and bead water. Under an electron microscope, oxidized paint looks like the floor of the Mojave Desert in the middle of summer: cracked, dry, and lumpy. Fresh paint, and paint that has been maintained with regular polishing and waxing, appears silky smooth and even.

Solutions that Shine

There are three ways to make the exterior of your automobile shine like new. First is a new paint job. A good paint job is expensive but may be your only option. This depends entirely on the paint's present condition. If it is oxidized to the point that primer shows through, paint it. Get a number of estimates and ask each painter to explain just exactly what you'll be getting.

The second way to make a paint job look new is to buff it or have it buffed by a professional detailer, who will use a buffing machine in conjunction with certain buffing pads and various polishes.

Your third choice is to polish and wax by hand. For oxidized paint, this is a labor-intensive operation. On the other hand, paint that has been moderately maintained should polish out quite easily.

Polishing removes the top layer of dead paint and exposes the good paint that was underneath. If oxidation is not too severe, you can save the paint job with polish and then protect it with wax. However, if polishing does not bring up the shine or you find yourself polishing into the primer, it is too late—the vehicle simply needs a new paint job.

TECHNIQUE 25 REMOVING SEVERE OXIDATION

Time: 1-1/2 to 4 hours

Tools: Polish, sponge applicator, clean soft cotton towels, cut-off paintbrush or soft toothbrush

Talent: ★✎

Tab: $5–$10

Tip: Ensure the ambient temperature is warm, at least 60 degrees Fahrenheit

Gain: Rejuvenated paint finish

Complementary project: Polish trim and brightwork

Two compounds are available: rubbing and polishing. These paste products are designed for hand application. They contain grit that removes dead paint, wipes out stubborn stains, and removes the toughest of paint blemishes. Except for polishing out small paint blemishes, they are used as a last resort before wet sanding or opting for a new paint job.

Since rubbing compound is so very strong and gritty, don't plan to use it until after a milder method has failed completely. Polishing compound is strong enough for most any job you should come across. Use polishing compound after you have tried a sealer/glaze polish. The objective is to renew paint while removing the least amount of good surface material. Polishing compound will remove more paint than sealer/glaze polishes.

Many auto enthusiasts have enjoyed very good results applying polishing compound with a small rectangular sponge that measures about 4 inches long and 3 inches wide. The sponge's straight sides offer users an easy guide for polishing along trim, molding, and other obstacles. Make sure the sponge is rinsed with clean water before using.

If a sealer/glaze polish doesn't work to remove oxidation and bring up a rich shine, apply polishing compound according to the instructions on the label. While rubbing on the body surface, use a back-and-forth motion, rather than a circular one. Circular movement causes swirls that are eliminated with the straight back-and-forth pattern. Do a small section at a time, about 2 square feet, for instance. Let it dry and buff with a clean soft cotton cloth. One application is generally enough, but don't be afraid to try again if the results are not satisfactory.

The same method works equally well to remove deep scratches that have not gone down far enough to reach the primer coat. Concentrate on the scratch and then work on the area around it to blend the surface and color together.

Polishing compound does not fill small cracks in newly exposed paint. Sealer/glazes fill cracks and scratches and add oils to paint to rejuvenate luster and shine. Therefore, you must plan to go over your vehicle a second and likely

Severely oxidized paint will likely need the strength of polishing compound. Seldom should you need the tremendous power in rubbing compound. After an application of polishing compound, you must follow up with a much milder sealer/glaze to remove swirls and then a coat of wax for protection.

Exterior Shine 83

a third time with a milder polishing sealer/glaze to bring out the paint's maximum shine and deep luster. These steps will also do well to remove minute scratches and swirls left behind by the abrasive nature of polishing compound.

After a section has been polished, glide the back of your hand across the surface. It should feel exceptionally smooth and sleek. If not, go over it again. As a side of your buffing cloth gets covered with dry polish residue, unfold to a clean side. Don't be surprised if you go through three or four towels. Use a small 1-inch-wide paintbrush with the bristles cut off to about 3/4 inch to loosen and whisk away dry polish from around trim, door handles, and the like.

1. The paint finish on this vehicle didn't dull overnight. It has been neglected for some time. After a wash, the paint on the fender will be treated to a couple of applications of polishing compound, prewax cleaner polish, a sealer/glaze, and then wax.

2. Polish is applied with a small rectangular sponge, about 3x4 inches. Application is made with a straight back-and-forth pattern to minimize swirls and spider webbing. Care was taken to avoid smearing the fender lip trim with polish by controlling the straight edge of the sponge applicator right next to and not on top of the trim piece. Notice the amount of red residue on the sponge. This is old, dead, oxidized paint removed from the surface.

3. The difference in this fender is amazing. Notice how the color and depth have improved dramatically. Polishing compound removed the bulk of oxidation and the other polishes worked to further shine the paint to bring out this rich finish. A couple of light applications of wax will protect this fender for three months or so.

4. From this angle, it is really easy to see the difference between the neglected paint on the hood and the polished paint on the fender. Not every neglected paint finish can be saved, as some will have oxidized so badly there is no good paint left under the dead paint to polish.

TECHNIQUE 26 CLAY BAR

Time: 1 to 2 hours

Tools: Clay bar, lubricant, soft cotton towels

Talent: ★♪

Tab: $15–$25

Tip: A lubricant must be used with the clay bar every time

Gain: Superbly cleaned and shined paint surface

Complementary project: Wax job

Many things can settle on an automobile's surface to make it rough to the touch, dull, and lackluster. They may include industrial fallout, acid rain, tree sap, paint overspray, bird droppings, brake dust, and airborne metal products. Among others, Meguiar's Quik Clay Detailing System and Mothers Clay Bar Paint Saving System are designed to shear off and remove such contaminants from your car's paint finish in preparation for polish and wax.

Used in conjunction with a lubricant, like Meguiar's Quik Detailer or Mothers Showtime Instant Detailer, the clay bar is gently rubbed across the surface in a straight back-and-forth pattern. It picks up and suspends contaminants to leave the paint finish smooth and clean. Many auto enthusiasts have rejoiced at the results they have found using nonabrasive clay bar products, which are advertised as safe for all paints.

According to instructions, make sure your vehicle is thoroughly washed and dried. Wet a 2-square-foot area with the liquid spray "detailer" product that comes with the bar. Instructions advise not to use water, although some detailers have noted that a liquid soap/water mixture can lubricate the surface adequately. Then gently rub the bar in a straight back-and-forth pattern across the paint finish. Maintain overlapping strokes using very light pressure; there is no need to rub hard. Apply more of the liquid detailer as needed to maintain wetness. As Mothers' instructions describe, "You will hear the clay bar removing the contamination. As the sound diminishes, lightly feel the area for any missed or stubborn contaminants." Continue with the procedure, ensuring that plenty of lubricant is in place, until the paint feels completely smooth. Wipe off excess liquid with a dry cloth and move on to the next section.

Once a body panel has been cleaned, fold the bar up into a ball and then flatten it out again. This will bring up a new section of the bar and keep the suspended contaminants trapped inside. Each palm-sized bar is advertised as being large enough to clean at least three average sized automobiles

To enhance the efforts of paint polishes, cosmetic car care manufacturers have developed "clay bars." These products are great for removing contaminants from paint finishes. They *must* be used with a lubricant. Mothers packages its clay bar with Showtime Instant Detailer for use as the lubricant and a small container of carnauba wax for lasting protection.

Exterior Shine

before it is too contaminated to perform as expected.

Once you have serviced your vehicle's paint finish with a clay bar treatment, be prepared to go over the surface again with a coat of sealer/glaze polish and then a light coat or two of carnauba-based wax. Meguiar's contends that after you have cleaned your car with its Quik Clay Detailing System, "... polish and wax will apply and wipe off in half the time, with dramatically improved results."

Not all clay bars are the same. Some come with different grades of polishing ability; medium and fine, for example. Read the directions on the label before "claying out" the finish on your vehicle. It is very important to follow instructions as to the proper lubricant to use.

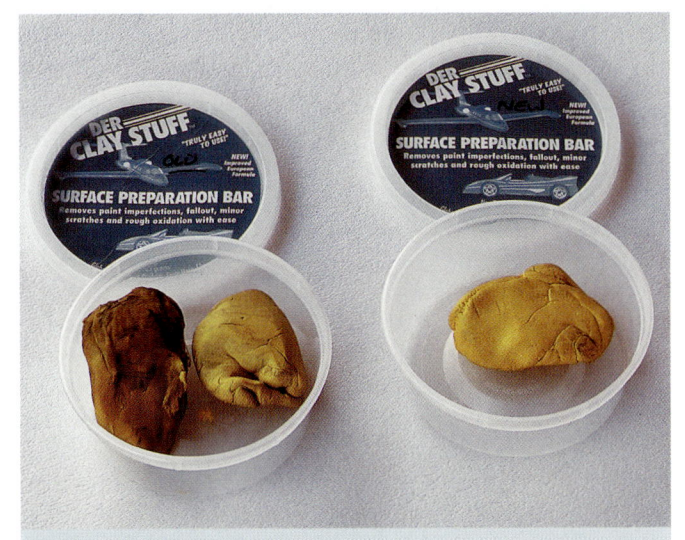

After you have finished using the clay bar, roll it up into a ball. This will fold under the part that has removed contamination from the painted surface to reveal a brand-new side ready for action. When you are ready to use it again, just flatten it out.

A clay bar is used to remove gasoline stains from under the fuel filler area. The spot has been lubricated and the bar is being gently glided across the surface, as there is no need to rub hard.

After the clay bar has done its job, the spot is polished with a mild product to remove any surface residue. A light coat of wax afterward will protect the newly polished area.

TECHNIQUE 27 BUFFING WITH A MACHINE

Time: 1/2 to 1-1/2 hours

Tools: Buffer, pads, polish, soft cotton towels

Talent: ★–★★★

Tab: $25–$50; more should you opt to purchase a buffer

Tip: Never allow a buffer to rest on one spot—keep it moving at all times

Gain: Rejuvenated paint surface

Complementary project: Do a wax job and clean the windows

High-speed buffers can be purchased for about $175 or rented at a tool rental yard for about $15 a day. A slow-speed Waxmaster model costs about $45. Regardless of whether you purchase or rent, you'll have to buy buffing pads. Meguiar's and other companies also offer orbital and dual-action buffers/polishers from $175 to $250. Some use soft cottonlike pads and others come with foam pads or cotton bonnets.

Before choosing a buffer, consider the rpm it delivers. Novices should start out with the low rpm orbital and dual-action models or lower rpm buffing machines. The slower, 1,700-rpm pad speeds are more forgiving than those professional models that produce up to 3,000 rpm. The slightest mistake using a 3,000-rpm buffer can result in paint burns, dislodged moldings, broken windshield wipers, or broken radio antennas. You'll have to decide if the potential saving of time is worth the risks of using a high-rpm professional buffer.

Two types of pads are available for most buffers/polishers: a cutting pad and a finishing pad. The cutting pad is used with fine sealer/glaze liquid polish to remove oxidation and scratches. Results, especially on dark colors, will look great but include fine swirls. After buffing with a cutting pad, you should buff again with the softer finishing pad and a nonabrasive liquid wax. The end result will be nearly perfect with no sign of swirls.

The polish you use with a buffer or polisher should state that it is recommended for machine application. Meguiar's Professional Machine Cleaner Number 1 and Machine Glaze Number 3 are specifically designed for machine use on paints with severe oxidation, scratches, water spots, and other serious paint problems. New Car Glaze Number 5 can be applied by machine or hand and is designed for paints with less than severe oxidation. The difference between them is the amount of grit contained in their formulas. It is usually best to start out with the mildest polish first and graduate to the more abrasive formula as necessary.

Wax for the finishing pad should also be in the liquid or cream form, and it should be made from a carnauba-based formula. The combination of soft, nonabrasive wax and a soft finishing pad removes swirls and leaves paint with a deep luster and mirror finish. For added protection, apply a final coat of carnauba wax by hand. The results should be stunning.

If you decide to use a Waxmaster Random Orbital, Meguiar's Dual-Action Polisher, or a similar machine, be certain to use their recommended polish and wax products to guard against damaging paint. Remember, it is much better to go over a paint surface two or three times with a mild polish than once with a more abrasive product that could result in paint burns.

Using the Buffer

If you have never used a buffer and are determined to learn how, practice on an old sled before tackling your classic show car. Immediately understand that if the buffer is improperly placed on top of a fender ridge or pressed too tightly in

The Waxmaster polisher is a slow-speed machine that employs a foam pad outfitted with a soft cotton bonnet for buffing. Notice how the detailer has the electrical cord draped over her shoulder. This is an easy way to keep cords from rubbing on top of fenders and from becoming caught up in buffing pads.

Right: This high-speed buffer is outfitted with a cutting pad. This pad is used with polish to remove oxidation and surface contaminants and shine paint. Use of this pad along with a polish designed for machine use will leave behind exceptionally light swirl marks generally only visible in direct sunlight.

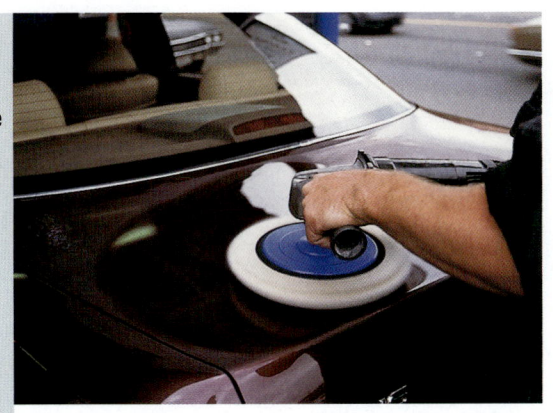

Far Right: A foam finishing pad is used on the buffer with a liquid or cream wax. This effort is not intended to shine paint. Rather, it will remove the minute swirls left behind by the cutting pad and polish.

Below Left: Start the buffing project by laying out a line of polish. With the buffer off, spread the polish with the pad motionless. This will greatly help to reduce the amount of polish splattered about after the machine is turned on.

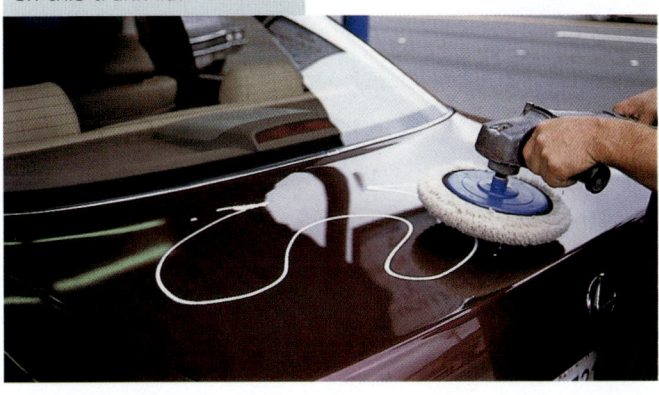

Below Right: Keep the buffer moving at all times. Never let it sit on one spot or rest on top of a ridge. If you do, you'll end up with a paint burn—the rapid removal of paint right down to the primer. Buff up to a ridge and never on top of it, as is being done on this trunk lid.

a door handle pocket, you will buff paint down to the primer in a split second. This is called a paint burn.

Polish, like Meguiar's Number 1, comes in a plastic bottle with a spout cap. Spread three lines of polish over a 2x2-foot section you intend to buff. You might find it easiest to start on the back half of the driver's side of the hood. Then move to the front half of the hood, front fender, roof, driver's door, and so on around the vehicle in a counterclockwise direction. This pattern works well for right-handed folks; try the opposite for lefties. Buff a section at a time, using the car's body lines, ridges, and trim as guides. Buff until the polish is gone and the paint shines; this will usually mean three to four passes over an entire 2x2-foot section.

Spread polish around the 2x2-foot section with the buffing pad before turning the machine on. This will help to minimize polish splatter. Keep the machine moving at all times. Allowing the pad to spin on one spot will surely result in a paint burn. Buff up to ridges and never on top of them. Stay away from body side moldings. If the pad catches on trim in just the right spot, it will tear it loose from the vehicle. The same is true for windshield wipers and emblems. Use extra caution around radio antennas. More than one detailer has been struck on the side of the face by the whipping action of an antenna that has been touched by a fast spinning pad. It is best to remove antennas, trim, emblems, and windshield wipers if possible.

Buff a section more than once when results are not satisfactory after the first attempt. This may be necessary when starting with a mild polish. If results are no better after a second pass, try a polish with more grit. This common occurrence is the reason why most professional detailers have an assortment of polishes on hand. Many have special blends they have experimented with over the years. Still, even professionals prefer to start with a mild polish, graduating to a grittier one as needed.

The electrical cord on a buffer/polisher is not generally much of an obstacle. You may notice a problem with it dragging on a fender while you buff the hood. In this case, drape the cord over your shoulder to keep it off the car. Also drape it over your shoulder while buffing the roof and trunk lid. Be careful while buffing lower body sections, as you do not want the cord to get caught up in the pad.

Once the buffing pad makes contact with the polish, you will quickly notice

Detailing Techniques

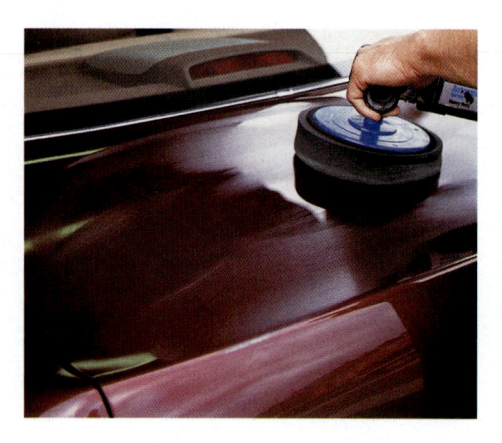

that polish tends to splatter. Even after initially spreading it around with the buffer off, you'll likely notice tiny spots of polish fly onto you and nearby objects. Be prepared. If you are buffing in the tight quarters of your garage, consider covering tools and other items within 6 feet of the vehicle. Think about wearing an apron and safety goggles to protect yourself.

Periodically, pads become caked with polish. If you're using a foam pad or terry cloth bonnet from an orbital or dual-action polisher, be sure to follow the instructions on how to clean them. For the thick pads used on higher-rpm buffers, use a spur or dull screwdriver for cleaning. Carefully lay the buffer upside down to rest on your knee, while maintaining a firm grasp on the handle. Reposition your hand on the handle so your thumb can activate the trigger. Turn on the machine while making sure your grip is still secure. Gently press the spur or dull screwdriver into the pad, starting at the outer edge and moving toward the center. You will see the color of the pad change from the color of your car (dead paint) to its natural off-white. Do this about three times to ensure the pad is clean. Plan to clean the pad three or four times before the job is completely finished; more if the vehicle suffers from severe oxidation.

Buffing the sides of a vehicle is the most difficult because of the awkward position you must work in. When bending down, you may be most comfortable if you reverse the position of your left hand, the one holding the extension handle off the side of the machine. Instead of grasping the handle with your thumb next to the buffing head, hold the handle with your little finger next to the head and your thumb away. Rest your left forearm on your left knee. This works especially well while buffing the lower sections of doors and quarter panels. Find a comfortable position for yourself, always keeping in mind the hazards when buffing near emblems and trim.

Once buffing is complete, your car should shine nicely. It will also be covered with fuzz from the pad and spots of splattered polish. This is quickly cleaned with a soft cotton towel or clean piece of flannel or terry cloth. Buffing with a finishing pad and wax will not be so messy, nor will polishing with a slower orbital or dual-action foam pad machine. Buffers/polishers are fast and they do a good job; simply hand waxing the entire vehicle still provides the easiest cleanup of all.

Buffing lower sections of automobiles requires a slight change in hand positioning. The left hand holding onto the side grip has been reversed. Instead of the thumb being next to the machine, the little finger is. Moving the left hand in this manner makes buffing lower body sections much easier and provides better control.

Above Left: Cutting pads have to be cleaned a few times during a buffing project. A pad spur is gently depressed into the pad after the buffer is turned on. The spur causes dry polish and contaminants to fly out of the pad. If you don't have a spur, you can use a dull screwdriver.

Above Middle: Buffing with a foam pad and wax is not much different than buffing with a cutting pad and polish. You can still burn paint, so be careful. Keep the buffer moving and clean the pad as necessary. Cream waxes work best and last longer than liquid wax products when used with finishing pads to buff paint.

Above Right: Whenever using a buffer, you have to pay attention to what you are doing! Getting too close to antennas will cause them to whip around and into your face. Getting on top of emblems may cause them to quickly break loose. High rpm buffers will quickly peel off body side molding, trim, badges, and almost anything else that is not firmly attached to the car body.

TECHNIQUE 28 HAND POLISH

Time: 1 to 2 hours
Tools: Polish, applicator, clean soft cotton towels
Talent: ★↗
Tab: $5–$10
Tip: Make sure the ambient temperature is at least 60 degrees, and work in the shade
Gain: Rejuvenated paint finish
Complementary project: Wax job

Special Products for High-Tech Paints

Clear coats and urethane paint finishes need special polish and wax. Meguiar's Professional Hi-Tech Cleaner Number 2 is designed to remove oxidation and harsh scratches and restore color on clear coats and urethane paint finishes. Since the instructions recommend you follow with Meguiar's Professional Hi-Tech Swirl Remover Number 9, you know that this cleaner is intended to remove heavy oxidation. Try the less abrasive swirl remover first to see how well it works. If it does the job, fine. If not, use the cleaner and follow up with the swirl remover to polish the finish to perfection.

Complete the job with a nonabrasive carnauba-based wax, like Meguiar's Hi-Tech Yellow Wax Number 26. The end result will be a deep paint finish as smooth as silk and as clear as glass.

Applicators

Polish manufacturers often recommend using small foam applicators for their products. These pads work fine. Another handy applicator is a small 3x4-inch rectangular sponge. These are commonly found in supermarkets in bags of four or more. They are handy in size and economical. Their straight edges allow controlled polish application around emblems, trim, rubber moldings, vinyl tops, and the like. As they get dirty, simply rinse with water. Throw away torn or damaged ones.

Use other applicators as you wish. Soft cotton towels, baby diapers, and old cotton T-shirts or flannel shirts work fine. Fold them into manageable sizes. As they become soiled, refold to a clean side. You might even consider going to the fabric store and purchasing a yard or two of soft cotton flannel. Cut it down into workable sizes, about as big as an ordinary hand towel. Cloths that are too big are cumbersome and hard to work with.

Putting polish on a car requires no great skill. There are some things to keep in mind, though. Remember, more is not necessarily better. Two light applications are better than one heavy dose. This is what "use sparingly" means on labels. Always try to employ the same front-to-back pattern whenever rubbing on paint. This reduces the chance of spider webbing and swirls.

When applying polish, slow down around emblems and trim. Polish smeared on these and other obstacles will make more work for you later during removal. The vehicle will not look as good as it should if this kind of polish residue is not removed. When applying polish with a cloth, use your finger to guide around emblems, antenna bases, and so on.

Sealer/Glaze Polishes

Polishing compound may be required for severe oxidation. Liquid polish relies on chemicals and very fine grit to remove light oxidation, enhance the

There are no big secrets behind polishing paint finishes by hand. Basically, be sure the paint surface is clean and dry. Then apply polish with a rectangular sponge or foam applicator in a straight back-and-forth pattern. Using a cloth to apply polish can be cumbersome, as indicated here.

shine, and add important oils to paint. Some products may include silicone additives that work well but could cause paint fisheye problems for future new paint repairs. Generally speaking, one-step polish/wax products contain silicones, and you must be aware that future paint touch-up or repair procedures will require fisheye preventive additives to prevent problems.

Moderate to light oxidation and swirled paint finishes should not require the strength of polishing compound. In those cases, use a sealer/glaze polish like Meguiar's Number 7. Its blend of rich polishing oils does an excellent job of shining good paint to perfection. Read the labels of the polishing products on the shelf of your local auto parts store to determine just what they are capable of accomplishing. Some are designed for moderate oxidation, while others are better for light swirls and spider webbing.

Labels on most polishes are quite specific as to what can be expected of the product. Some manufacturers, like Meguiar's and Mothers, have even produced series of products intended for use as full packages. For example, Mothers offers a three-phase "Ultimate Wax System" that includes its Pre-Wax Cleaner as phase one for removing oxidation, Sealer and Glaze as phase two for hiding swirls, and Pure Carnauba Wax as phase three for long-lasting shine protection.

The Mothers three-phase system is designed so that the phase one Pre-Wax Cleaner should only be needed once or twice a year. The phase two Sealer and Glaze and the phase three Pure Carnauba Wax can be used as often as desired to maintain the vehicle's perfect shine.

It is virtually impossible to recommend one polishing product over another, as each vehicle presents its detailer with different types of paint finish, oxidation, scratch, and swirl problems. The only way to truly determine which product should work best for you is to read labels. Automotive paint care manufacturers like Meguiar's, Mothers, 3M, and the rest have done a lot of research, and they provide you with clear explanations on their products' labels as to what they have found their products are capable of accomplishing.

Application

Your vehicle should be clean and dry before starting any polish job. Complete the work in the shade. If your work area is such that only a part of the car can be

Removing dry polish simply requires a clean soft cotton towel or cloth and some elbow grease. It is best to polish just in small sections, about 2 square feet, and then wipe off with a towel when dry. If you notice that sealer/glaze is very difficult to remove when dry, it may be because the surface is badly oxidized and in need of a more potent product, like prewax cleaner or even polishing compound.

in the shade at one time, plan to polish the shady part first and then reposition it so that you can work on the other side in the shade. Let the heated side cool off before applying polish. This holds true for wax jobs, too.

Apply polish with a damp sponge, soft cloth, or foam applicator. Start at a specific spot and work your way around the vehicle in a systematic manner. Generally speaking, a good place to start is at the back half of the hood on the driver's side. Spread polish on your applicator and use a back-and-forth pattern to cover about 2 to 3 square feet, or about one-quarter of the hood. Try to avoid smearing polish all over trim, windshield washer nozzles, and other obstacles.

After a few passes, or as the polish begins to dry out, stop for a moment to allow the rest of the polish to dry. Use a soft, clean cotton cloth to wipe off the dry residue, remembering to turn your cloth over to a clean side frequently for best results.

Once the driver's side of the hood is complete, work on the front driver's side fender. Then move to the driver's side of the roof and the driver's side door and work toward the trunk. Polish the farthest-to-reach and highest points first, like the hood and roof, before tackling the lower parts. This way you can avoid having to lean against freshly polished sides of the vehicle.

You will need more than one clean cloth to remove polish. As polish, oxidation, and dead paint residue accumulate on the buffing cloth, you'll find that it becomes harder and harder to remove the dried polish. Plan to unfold the cloth to a clean side after each 2-to-3-foot section is buffed clean. Continue the process until you have gone around the vehicle completely in a counterclockwise direction to finish up at the passenger side of the hood.

Don't spend too much time attempting to remove every speck of dried polish from around emblems, badges, trim, and the like. Do a good job, of course, but don't fret the tiny stuff. You will have to go over the entire vehicle again with wax. Once that job is done, you will want some quality time to tidy up and focus on every fleck of dried polish and wax.

TECHNIQUE 29 HAND WAX

Time: 1 to 2 hours

Tools: Wax, applicator, clean soft cotton towels

Talent: ★★

Tab: $10–$20

Tip: Apply wax sparingly, as a little goes a long way

Gain: Protection of the paint finish

Complementary project: Clean the windows

Good paint needs wax to help keep it fresh, seal in the protective polish oils, and prevent the sun's ultraviolet rays from drying it out and causing oxidation. Paint also needs to breathe, and wax helps to keep its microscopic pores open. Paint was originally put on cars to keep metal from rusting. Wax preserves paint and thus helps to prevent rust (oxidation).

It is not easy to recommend one brand of wax over another. Most car people agree that any product that highlights carnauba wax as the base ingredient is good. Meguiar's, Mothers, Eagle 1, and others clearly note on product labels that their wax is specifically intended for use on cleaned and polished surfaces. They also recommend that wax be applied sparingly and advise that a little wax will go a long way. This means that two light coats of wax are better than a single heavy application.

The lasting quality of wax varies with climate, sun exposure, and temperature. By far, carnauba-based waxes last the longest, generally about three to six months, depending on such factors as weather, parking conditions, car cover use, maintenance, and car wash soap. During summer months, the roof, trunk lid, and hood need the most protection. This is due to the penetrating rays of the sun and high temperatures. In winter, the sides take a beating from road grime, sand, and salt. It is a good idea to keep this in mind and concentrate more frequent waxing on those areas affected during each season.

Carnauba-based waxes may be difficult to buff off during cold weather and may leave streaks and clouds. You will have to experiment with them. Ideally, wax your car in a heated garage or shop to minimize problems. If not, you may have to settle for a one-step cleaner wax during those times.

Liquid waxes containing carnauba generally last as long as the pastes or creams. Spray waxes may not last as long because the contents must be thinned to allow the solution to flow out of the tiny nozzle opening.

Automobiles should be waxed four times a year. A cleaner polish is recommended once or twice a year to remove old and yellowing wax. Newer cars with high-tech paint jobs should be waxed with products especially designed for them.

Water will bead up quickly on a waxed surface; on a surface with little or no wax, it will just sort of sit there. When water fails to form into tight beads on the surface, you know it is time for a fresh coat of wax. A little goes a long way, so plan for two light coats rather than a single heavy dose.

Exterior Shine

Right: Wax is applied with a clean, small rectangular sponge or foam applicator in a straight back-and-forth pattern. It is important to use a clean soft cotton towel or cloth for dry wax removal. Along with soft white terry towels, Meguiar's offers Ultimate Wipes, specially made cloths designed for removing dry polish and wax. Each cloth can be laundered up to 500 times.

Far Right: You should not polish over pinstripes or painted graphics. Most of the time, these artist's works are completed with paint different than that on your vehicle's surface. Although you must not polish these items, you can apply wax over them after they have fully cured, usually 30 to 60 days after they have been painted on.

The ultra polish/wax job for most any automobile would be one or two applications of a cleaner polish, one or two applications of a sealer/glaze, and two coats of a carnauba-based wax. It may take a couple of days to accomplish such an undertaking but the results will be brilliant and last for months.

Apply wax much as you would polish. Start at the hood, work down the front fender, over to the roof, down the door, and back toward the trunk, the rear quarter panel, and the rear of the vehicle, until you get back to the hood. Directions on containers of wax recommend you do small 2x3-foot sections at a time. Buff off wax as it dries. Some detailers like to cover entire vehicles with wax in one step and then buff. The choice is up to you, as long as the vehicle is totally in the shade. Do not wax your car in the sun.

While buffing off wax, turn over your soft terry cloth, cotton, or flannel cloth frequently to a clean side. As the cloth picks up wax residue, its ability to pick up more is extremely limited. You'll likely find that it may require at least two cleaning cloths to sufficiently buff off the fresh wax from an entire vehicle.

If you intend to apply a second light coat of wax, do not spend too much time trying to remove every hint of wax residue from around emblems, badges, and the like. You can do that later, after the second coat has been buffed off. There are a few tricks to accomplishing the detailing chore of in-depth polish and wax removal, and they'll be covered in the next project.

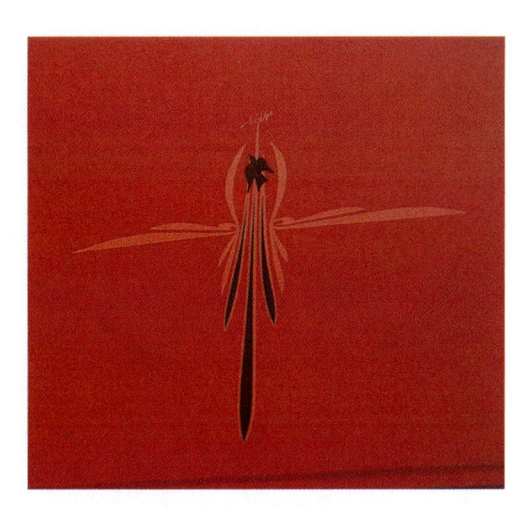

Pinstripes and Racing Stripes

Painted racing stripes and pinstripes are cared for the same way you wash, polish, and wax body paint. Vinyl tape stripes may be waxed along with paint. Be sure to apply wax in the same direction as the stripe to minimize wax buildup along the stripe's edges. Residual dry wax is removed from edges with the cut-off paintbrush or your fingernail inserted inside a soft cloth.

Pinstripes are generally done with paint products used for sign painting. Not nearly as "bulletproof" as automotive urethanes and other high-tech paints, these products nevertheless hold up quite well. It is perfectly all right to cover them with wax while waxing your car. However, you must wait until paint has fully cured before waxing over brand-new pinstripes, as your pinstriper will undoubtedly tell you.

Waxing New Paint

Paint needs to breathe. This is especially important for new paint, in which solvents and other additives need time to fully cure. If your vehicle has recently been painted, the autobody paint specialist should have advised you to refrain from waxing for as long as 90 days. This is to ensure that all of the solvents and chemicals in the paint have plenty of time to fully cure and evaporate from the surface. Waxing new paint too soon could result in paint flaws.

TECHNIQUE 30 REMOVING POLISH AND WAX RESIDUE

Time: 1/2 to 1 hour

Tools: Cut-off paintbrush, toothbrush, clean soft cotton towels

Talent: ★♪

Tab: $5

Tip: When finished, pull vehicle out into sunlight for final inspection

Gain: Crisp-looking paint finish

Complementary project: Dress the tires

Just about everybody knows how to remove wax once it has dried on the paint finish. But many forget the details and leave behind remnants of polish and wax in all sorts of little nooks and crannies. Spots like these can take away from an otherwise crisp-looking machine. Along with removing the big stuff after a polish and wax job, you must take the time to remove all traces of polish and wax from the little spots.

Use a soft cotton cloth to wipe off dry polish or wax, unfolding regularly to a clean side as residue builds up. The cloth you use should be very soft and absolutely nonabrasive. Car enthusiasts use baby diapers, cheesecloth, discarded T-shirts, bath towels, and plain cotton flannel purchased at fabric stores. Many of the name-brand cosmetic car care product manufacturers also make polish- and wax-removing cloths available.

Whichever cloth you employ, cut it into manageable 2-foot squares (if need be) and fold it into quarters. Use one side to remove the bulk of polish or wax and follow with a clean side to pick up anything left over. Do not try to get more out of a side than practical. As the cloth soils, just open it up to a clean side and refold. Eventually, you'll need to retrieve a new, clean cloth. Be prepared to use at least two for each polish and wax removal exercise. Employ a straight back-and-forth pattern as you move around your vehicle removing dried polish and wax.

A 1-inch-wide paintbrush with the bristles cut to about 1/2 to 3/4 of an inch works great for removing polish and wax buildup from emblems, mirrors, light lenses, louvers, antenna bases, and the like. Use it as you go about buffing off polish and wax. The stubby bristles have the strength to remove even the most stubborn wax buildup. Place a piece of duct tape around the metal band on the paintbrush that holds the bristles in place. This will help to prevent accidental scratches and paint chips while you are busy removing polish and wax buildup in tight spaces.

Start at the top of the vehicle and work down. This way, dust and powdery polish/wax residue will fall on parts of the vehicle you have yet to buff. Use the cut-off paintbrush around sunroofs, along window trim, inside door handles, in the lettering on plastic light lenses, and anywhere else you notice even a hint of polish or wax buildup.

After you have polished and waxed your car or truck, pull it out into the sunlight and inspect for any lingering traces of dry polish or wax. If you have just purchased the vehicle, there may be many places smeared with this residue. Use a soft cloth to wipe away the big stuff, using a fold to get into grooves and tight spaces.

Above Left: A 1- to 2-inch-wide paintbrush with the bristles cut to about 1/2 to 3/4 of an inch works great for breaking loose dry polish and wax from around emblems and badges, along trim, and inside recesses. Use it along with a soft cloth to remove every hint of polish or wax residue from your car's surface.

Above Right: After going over the vehicle a couple of times to find the obvious polish and wax buildup, take a few minutes to closely inspect the small stuff. Look along moldings, weather stripping, and body side trim. The better job you do removing every little trace of polish and wax, the better and crisper your car or truck will look.

Periodically, place your head close to the body surface and look at it from front to back or vice versa. Look toward the garage door opening or the place with the most sunlight or illumination. The reflection image you see in the paint should clearly show you where the missed spots are. Go back over them with a clean side of the buffing cloth to remove dried material.

If polish/wax residue dust lingers on the surface of your car, simply use a clean cloth to lightly wipe it away. Polish and wax dust is very light and will settle back on a car after it has been disturbed and fluffed into the air during your buffing procedure. This is normal.

Go over the vehicle at least a couple of times with your attention focused on finding the slightest hint of polish or wax buildup in those hard-to-reach crevices and small pockets around license plates, back-up lights, body side molding, badges, hood and trunk edges, door frames, and the like. The better job you do of removing every speck of dried polish and wax, the better your automobile will look.

Polish or wax smeared on vinyl or rubber trim or moldings may be tough to remove. Dab a little cleaner on a cloth and rub. It that doesn't work, try using a clean toothbrush with cleaner. Once the spot has been either removed or dulled as much as it is going to get, treat that piece of trim or molding to a light coat of dressing.

Common spots to find polish and wax buildup residue are inside letters and emblems. Just a few minutes with a cut-off paintbrush and a soft cloth is all it will take to make these letters look crisp and clean.

96 *Detailing Techniques*

TECHNIQUE 31 PAINT BLEMISHES AND CHIPS

Time: 1/2 hour

Tools: Paint chip repair kit, paint, artist's fine paintbrush, wax and grease remover

Talent: ★★

Tab: $5–$15

Tip: Do not wax over paint chip repairs for at least 30 days

Gain: Improved paint finish appearance

Complementary project: Clean the tires and wheels

The best way to protect paint is with a good coat of wax. The rule of thumb is to wax when water stops beading on the surface. That rule is OK, but many believe waiting for water to stop beading is too late. They prefer to wax when beads of water start to flatten out, losing the firm round shape that they maintain on a freshly waxed surface. Others prefer to wax once a month without regard to water beads.

Paint Blemishes

Bird droppings, splattered insects, road tar, and tree sap all blemish paint quickly. A good coat of wax helps to protect paint, but removing these hazards as soon as possible is the best way to prevent stain blemishes. Cold water and a soft towel should be sufficient to remove most of these problems. Those that are more difficult to remove may require the use of a mild polish. You may find that one-step cleaner waxes also work well.

After a road trip, you may notice all kinds of stuff stuck to the paint on your automobile. A good wash job is certainly in order, but it may take a bit more than soap and water to remove some caked-on problems. Automotive cosmetic care product manufacturers produce a number of bug and tar removers, available at auto parts stores. Give them a try to see which works best for you. By all means, stay at it until the material is removed. Left in place, most of the things just described could eventually bond to paint and cause etching problems or permanent stains. On extended trips, consider carrying a small bottle of bug and tar remover and a rag in the trunk or glove box so that you can take care of such problems early on.

Paint Chips

Automobiles that are actually driven are subject to paint chips—there is no way to avoid them. You might apply a clear plastic guard around the bottom of the rear quarter panel and along the rocker panels and a bra across the front, but your ride is very likely to suffer a paint chip somewhere at some time.

Minor paint chips are repaired using touch-up paint. Most touch-up colors are available at auto parts stores, dealerships, and most autobody paint and supply houses. If a stock touch-up paint bottle is not available for the paint job on your vehicle, an autobody paint and supply store should be able to mix up a pint or quart to match the stock color of your car by using the color code from the vehicle identification number. They can also mix custom paints as needed.

Old-school car buffs used the bottom end of a cardboard matchstick to apply touch-up paint. You may have better luck using a fine artist's paintbrush, the brush attached

Included here are a paint chip kit and other paint blemish repair materials from Pro Motorcar Products. Some of the tools are used to clean out debris from paint chips in preparation for the application of new paint.

Exterior Shine

to the cap of touch-up paint you bought at the dealership, or the brush supplied with a paint chip repair kit found in some auto parts houses and through catalog sales. Paint chip repair kits are available from the Pro Motorcar Company and from The Eastwood Company for around $25 to $45.

Be sure the base of the paint chip is free from rust and other debris. Clean it with a cotton swab dabbed in wax and grease remover. Plan to apply more than one coat of paint. Let the first coat dry sufficiently before applying another. This may take a day or two. The new paint on the chip area will likely build up ever so slightly higher than the surrounding paint finish; this is okay.

When the paint has dried according to the label (generally about a week), mask it off and wet sand with very fine grit wet-and-dry sandpaper (number 600 or higher). Make sure you use plenty of water and sand lightly. Masking tape around the chip will protect the surrounding finish. When the bulk of paint has been reduced to the level of the surrounding finish, remove the masking tape and follow up with a couple of light applications of sealer/glaze polish. Don't worry about waxing for a month or so. This will give the paint time to cure.

1. A very slight scratch can be noticed midway between the window trim and the corner of the quarter panel edge. Polishing has not repaired the blemish, so the detailer will resort to wet sanding.

2. With plenty of water from the wash mitt, the detailer gently glides the ultrafine-grit wet/dry sandpaper over the scratch. A few swipes with the sandpaper are followed with a few swipes with the wet wash mitt. The mitt cleans off sanding residue and lubricates with each pass.

3. The area that has been wet and sanded is noticeably scuffed. The scratch is gone, and now it will take a little polishing to remove the scuff mark to make the paint shine.

4. Polish was applied, wiped off, and applied again several times to result in a perfect paint blemish repair.

Chapter 6
Underbody Detailing

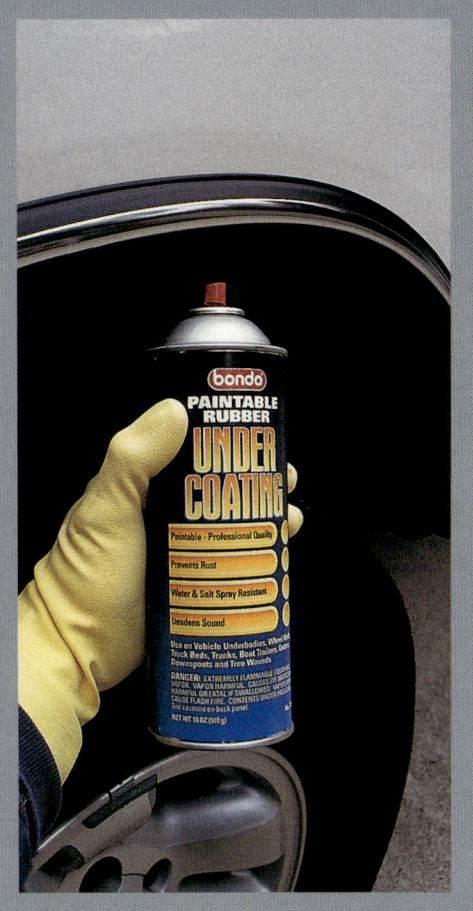

Looking under a car tells you a lot about its owner. What kind of car person are you? Put these underbody detailing techniques to good use.

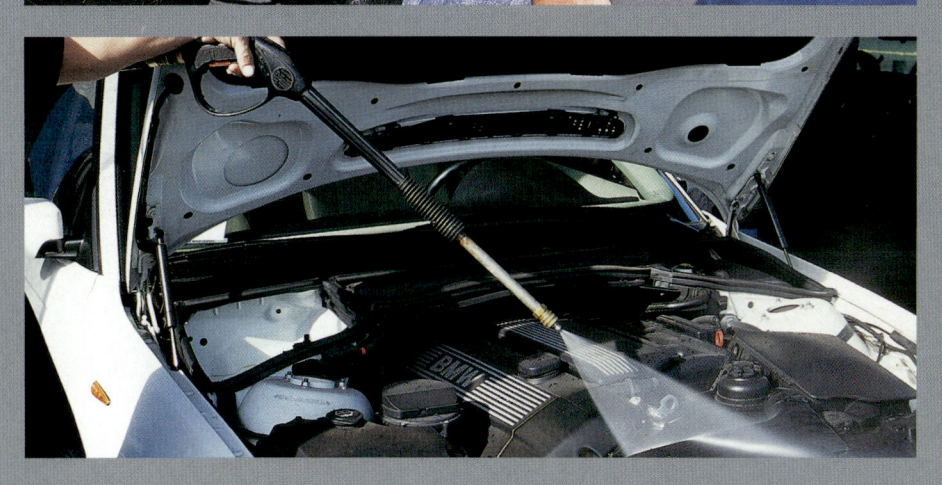

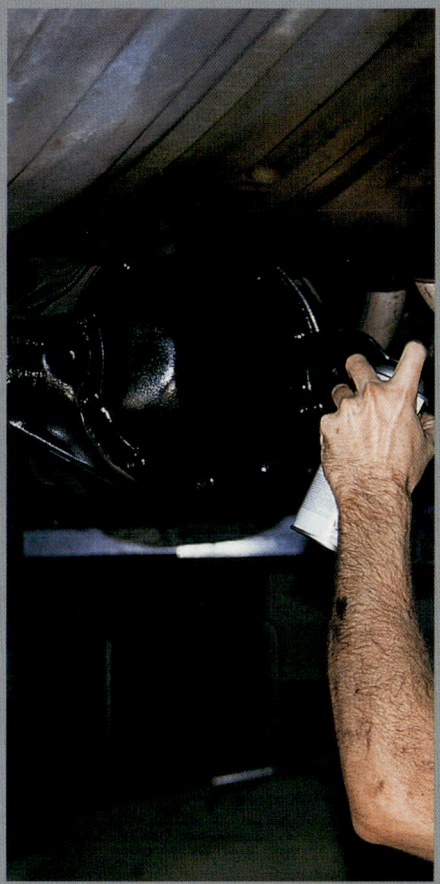

100 *Detailing Techniques*

Next time you drive by a used car lot, look at the fenderwells and visible underbody of cars on the front line. Do they look dirty? Does the back of the front fenderwell complement the wheel and tire? Are visible frame members under doors in good-looking condition? Cleaning and painting these areas helps to freshen the overall automobile appearance.

Squat down and look at your car from ground level. Notice the parts that catch your eye. The frame, exhaust pipes, mufflers, suspension, and fenderwells that are unsightly can draw your eye away from the rest of the car. Clean, paint, and shine those items so they add to the overall appearance of your otherwise fine vehicle.

Be certain that the place where you expect to clean and detail an underbody is right for the job. A driveway may not be suitable for underbody detailing on neglected vehicles; think about a steam cleaner facility first for the big stuff. When working under your car or truck, be certain it is supported with sturdy ramps or jack stands and that the tires on the downhill side are blocked.

Not many people will see this part of your car. However, just how many parts of the underbody are visible to other drivers? Drivers at stop signs can clearly see the areas below front and rear bumpers and the side frames below the doors. Various makes and models feature exposed underbody parts to different degrees. You need to squat down and take a look at your ride to determine just what needs to be detailed and what can be left alone.

Steam cleaning facilities are equipped to handle the greasy and grimy water runoff from cleaning underbodies and engine compartments. They will utilize lifts or pits to accommodate undercarriage work. Be aware that steam cleaning could result in damage to factory stickers.

TECHNIQUE 32 STEAM CLEANING

Time: 1 hour

Tools: Steam cleaner, safety goggles, rubber gloves, rain gear

Talent: ★★★

Tab: $45-plus, depending upon what extras you request

Tip: Ask to look at vehicles previously steam cleaned to inspect the quality of workmanship

Gain: Removal of crud and eyesores from undercarriage

Complementary project: Paint or undercoat steam-cleaned chassis areas

Most detail shops do not advertise steam cleaning abilities. Look for detailers with steam cleaning facilities in the Yellow Pages of your telephone book.

Expect to pay from $45 to $65 to have the undercarriage of your vehicle steam cleaned (detailers will clean the bottom half of the engine, the transmission, chassis, and fenderwells). For an added $35 to $45, they will also clean the top half of the engine and the engine compartment. Afterward, you must repaint the fenderwells and those frame parts that have suffered peeled paint or other finish problems.

Steam cleaners work faster than pressure washers because of the heat and high pressure they generate. Access to the chassis is made easy by the use of a lift. Care must be exercised to ensure that wires and connections along the frame are not compromised.

If your vehicle exhibits a filthy underbody and you are determined to get it clean, seriously consider the services of a steam cleaner. You will not have to contend with the mess and will save a great deal of cleaning time. Afterward, you can block the vehicle up on jack stands and paint or undercoat the underbody and fenderwells at home.

Detailers that offer steam cleaning services generally also provide undercoating work. Expect to pay around $125 for a complete steam cleaning and undercoating job. Remember that undercoating and rustproofing are completely different processes that employ vastly different products.

Undercoating material is designed to cut down on road noise and protect underbody sections from hazards like rock chips. Rustproofing materials are designed to protect metal from road salt and other rust-causing hazards. Although both begin with steam cleaning the undercarriage, rustproofing requires a much more thorough cleaning because the materials must bond directly to metal.

Undercoating will eventually dry to a relatively hard surface. Rustproofing materials tend to stay much more pliable. Undercoating is simply sprayed onto surfaces under vehicles, while rustproofing materials are also sprayed inside body cavities, including doors, quarter panels, inner fenders, and the like. Rustproofing is self-healing, meaning that a minor scratch on the material's surface will be re-covered by adjacent material. Undercoating will not self-heal.

If you choose to steam clean the underside of your vehicle, check the Yellow Pages for pressure washer dealers that also sell steam cleaners, or check to see if local tool rental yards have steam cleaning machines. Make sure you can secure your vehicle safely up off the ground and that you are ready for the greasy runoff. Be prepared to repaint, undercoat, or rustproof any metal sections that were cleaned down to bare metal, as soon as possible, to prevent the formation of rust deposits.

If you opt to undercoat or rustproof, check with auto parts stores and autobody paint and supply house to see if they carry the right materials. The Eastwood Company catalog offers both undercoating and rustproofing products, and the necessary tools. Expect to pay $10 to $50 for materials, depending on the size of the job you have on hand.

TECHNIQUE 33 THOROUGH RINSING AND CLEANING

Time: 1 to 4 hours

Tools: Water, soap, cleaner, brushes, dull putty knife, rubber gloves, safety goggles

Talent: ★★★

Tab: $10–$25

Tip: Secure raised vehicles safely on good jack stands

Gain: Removal of potential rust-causing problems

Complementary project: Repaint, undercoat, or rustproof cleaned parts

Fenderwells are constantly bombarded by road debris thrown up by tires. During winter driving, salt, sand, and mud are splattered throughout fenderwells and open areas close by. These include steering and suspension parts, wheel hubs, frames, and bumpers. Pockets of grime will retain moisture and give rust a start. This is why body cancer generally starts at the rear bottom of quarter panels, just behind fenderwells.

Thorough rinsing is not just a quick spray with a garden hose. It is a concentrated effort to wash away every loose bit of grit, sand, salt, and road grime. The most effective method requires raising a vehicle, bracing it with jack stands, pulling wheels, and flushing completely. You should do this at the end of each winter driving period. This is also a great time to wash the wheels, front and back.

The inside lips of fenderwell edges are prime spots for dirt buildup. Feel for this with your fingers, break loose with a brush, and flush with water. The top of the front wheel support bracket is another likely place to find dirt and grime buildup. Inside bumpers and on top of axles and frame are also areas that generally exhibit buildup.

Cleaning

Normally, garden hose pressure through a nozzle is sufficient. A pressure washer is better, although you must spray cautiously to avoid paint and undercoating damage. A plastic-bristled brush does a good job of removing stubborn buildup, especially when combined with a cleaner like Simple Green. This brush can be used on the inside of the fender lips without much risk of scratching exterior paint. Use a toothbrush to clean slots on screws attaching trim to fender lips.

The visible frame under doors should also be cleaned with a brush and soap and water. Set up a workable pattern for cleaning these areas on your car to minimize the amount of water runoff on the ground. Start the project at the downhill end of the work area and work uphill. This will help to keep you dry during the cleaning process.

The underbody below the rear bumpers is another spot that needs cleaning. Corvette designs allow that part of the car to be most visible. Use an old wash mitt and scrub brush to remove dirt and unsightly grime. This

The fenderwell area on this kit car is easy to access for cleaning and detailing. If the fenderwells on your vehicle are in dire need of detailing, consider pulling off the wheels for best access. Be sure the car or truck is securely supported on jack stands.

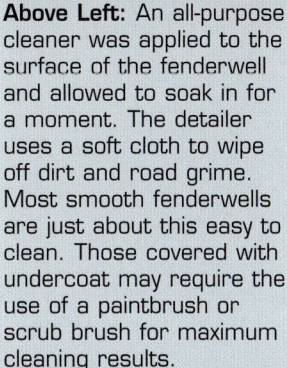

Above Left: An all-purpose cleaner was applied to the surface of the fenderwell and allowed to soak in for a moment. The detailer uses a soft cloth to wipe off dirt and road grime. Most smooth fenderwells are just about this easy to clean. Those covered with undercoat may require the use of a paintbrush or scrub brush for maximum cleaning results.

Above Right: Fenderwells this clean complement the rest of the vehicle. Nicely detailed tires and wheels are highlighted when surrounding fenderwell surfaces are also looking crisp and clean.

includes mufflers, visible shocks, and axle housing.

A once-a-year or first-time cleaning may require more than simple brushing. Dirt on A-frames and other frame members can cake up to an inch high. Carefully use a dull putty knife to remove the heavy stuff and do a lot of brushing with cleaner and water to remove the grime and get down to clean metal.

Clean everything inside the fenderwell. Brake lines, bleeder valves, light buckets, and wiring should come clean with soap and water. Use a toothbrush in tight spots and a plastic-bristled brush on flat surfaces. Slight rust formations are cleaned with steel wool and WD-40. Along with rust prevention, cleaning prepares surfaces for paint.

As you come across stubborn spots of dirt and grime buildup, spray them directly with an all-purpose cleaner like Simple Green and scrub with a plastic-bristled brush. Rinse with clear water and inspect the area to ensure that it is as clean as you want it.

The Process

Rinse the underbody at a site away from where you plan to actually wash it. This way, the major stuff will be removed and your in-depth cleaning area will remain dry. Let the undercarriage drip dry for a little while so your actual work area stays dry.

With the vehicle relocated to your driveway or other worksite and facing uphill, start cleaning at the rear underbody. Wear goggles while under the car to protect your eyes from falling debris. Whisk away as much of the loose stuff as you can with a dry scrub brush. Then spray all-purpose cleaner on the dirty spots and pick up a load of suds from the wash bucket. Scrub a small section at a time to ensure quality cleaning. Move onto another section while under that part of the vehicle; don't worry about rinsing with water until the entire area has been cleaned.

Assured that you have cleaned the area as well as possible, get the garden hose and rinse. You can kneel on the hose with one knee in an effort to keep your pants dry. Once the rear underbody is clean, move on to one side of the vehicle and repeat the process. Next will be the other side underbody, and last will be the front.

By the time you are finished, all parts of the visible underbody should be clean. Get on a creeper and roll under the rear, side, and front underbody areas to inspect them. Use an old wash mitt or cleaning cloth to pick up what has been missed. Use a droplight to aid in your inspection.

104   Detailing Techniques

TECHNIQUE 34 FENDERWELL PAINTING OR UNDERCOATING

Time: 1 to 2 hours

Tools: Paint or undercoating, masking paper and tape, paint thinner for cleanup

Talent: ★★★

Tab: $10–$25

Tip: Be sure areas to be painted are dry before starting

Gain: Crisp-looking fenderwells and visible underbody members

Complementary project: Polish chrome exhaust tips

Masking before Painting

Front fenderwells vary with each automobile. Some include open areas to the engine compartment, exposing the block and exhaust manifold. Others are fairly tight, exposing only struts and steering assemblies. All have brake, fender lip, and hub parts that will need masking.

Front end and steering assemblies, as well as drum brake housings, can be painted semigloss black. Two or three coats is sufficient. Mask lugs, brake fluid bleeder valves, grease fittings, brake lines, and everything you don't want painted. Three-quarter-inch masking tape is a good workable size, or you can purchase tape up to 2 inches wide. Use masking paper or at least two sheets of newspaper together for large masking jobs around wheel hubs and openings to the engine.

Chrome-trimmed and painted fender lips should be masked, even though controlled painting should prevent overspray. Wide masking tape covers these thin surfaces with one pass. You can also try cutting masking paper into thinner strips for more manageable masking. Remember to mask mud flaps, too.

Black fenderwells look good compared to rust-colored ones. The real difference is noticed in detail work. To break the monotony of solid black, allow grease fittings, brake connections, and the like to remain unpainted; it adds color to the area.

Painting

The type of paint to use on fenderwells is debatable. You can find paint products specifically designed for these vehicle areas through The Eastwood Company and most autobody paint and supply stores. Otherwise, plan to use a semigloss with rust-inhibiting agents. Of equal importance is surface cleanliness. Paint will not adhere to grease. If dirt is painted over, when the dirt vibrates

This fenderwell definitely complements the tire and wheel, as well as the rest of this classic automobile. The paint was applied to a clean surface and a number of coats were sprayed on, rather than a single heavy dose. Note the uniformity of the paint finish.

Undercoating is not rustproofing. Should you decide to apply undercoat, remember that its main quality is the lessening of road noise. Its texture is bumpy, not smooth. Be sure the fenderwell surface is clean and dry before application.

loose, it will take the paint with it.

It is preferable to paint in warmer temperatures. Heating a spray can in a sink of warm water expands the propellant and helps to better mix the binders and solvents in the mixture. This helps paint to go farther and flow on more smoothly. Never heat a spray paint can in a pot of water on the stove! The added heat can cause the propellant to expand too much, bursting the can and possibly resulting in a fire or injury.

Since most spray paint cans carry a warning not to expose can or contents to temperatures over 120 degrees Fahrenheit, hot water should be from the tap only. At no time should the water be too hot for your hand. The objective is to warm the paint, as it flows and adheres much better at 80 to 90 degrees Fahrenheit than at 50 or 60 degrees.

Follow the directions on the can and spray with even strokes. Continue with additional coats until you reach the desired effect. Always let the first coat dry before applying the next. Remove accidental overspray with glaze polish on painted surfaces and with lacquer thinner on others.

Undercoating

Some car people prefer to use undercoating on fenderwells instead of paint. Paint looks good on front-end and suspension parts, but flat on previously undercoated fenderwells. You will need two spray cans of undercoat to adequately undercoat all four fenderwells.

Heat the cans first in a sink of warm water for a few minutes. Apply undercoating as you would spray paint. You will notice the spray immediately bubble once it touches the fenderwell surface. If you apply evenly, the undercoating can make fenderwells look new.

Undercoating is not a rustproofing agent. It is intended to deaden road noise. True rustproofing requires a much more detailed process. Parts have to be completely cleaned down to bare metal and properly prepared with primers. Afterward, a waxlike rustproofing material that never seems to dry or harden is applied.

Combating Rust

Paint was originally put on cars to keep them from rusting. Salt is a predominant factor, as evidenced by severe corrosion problems on cars driven on salted roads and those in close proximity to bodies of salt water. Pockets of dirt, leaves, and other debris also retain moisture and should be removed from vehicles as soon as possible. Look for rust wherever moisture collects: floorboards, trunk, lower door sections, rocker panels, fenders, and battery boxes.

Once you notice rust coming through paint, it is too late. That section of the body will have to be repaired by a body shop. The best way to combat rust is by not giving it a chance to start. Frequent and thorough washing is a good beginning.

While cleaning the fenderwells and underbody, closely check seams and joints for evidence of rust. Remove surface rust with a soft wire brush or sandpaper. You can also use acid-based chemicals, like The Eastwood Company's "Rust Encapsulator," to kill rust. Be sure to follow the instructions carefully. After you have removed the hazard, dry the area completely and cover with two coats of good primer. Afterward, cover the area with a couple of coats of semi-gloss paint or undercoating.

Rustproofing products are available at most auto parts stores. Sealant sprays come in aerosol cans. They are equipped with small hoses designed for application through drain holes in doors or other holes allowing access to open cavities. The entire area must be coated and any holes made for access must be covered. The sealant's main purpose is to prevent moisture from coming in contact with bare metal.

Undercoating is not designed as a rustproofing agent. If water is trapped between it and the metal, rust will occur. If it peels loose or chips off, moisture will come in contact with the metal to pose new rust problems.

DETAILING Techniques

TECHNIQUE 35 FRAME MEMBERS AND TAILPIPES

Time: 1/2 to 1 hour

Tools: Paint, paint block, masking tape, and paper

Talent: ★★✦

Tab: $10–$20

Tip: Ensure parts are clean and dry prior to painting

Gain: Eliminates eyesores to make members almost invisible

Complementary project: Dress the tires

Painting Visible Frame Members

Most of the time, it is quite easy to paint visible frame members using a paint block instead of masking. A paint block is nothing more than a light piece of cardboard about the size of a license plate. You will have to go slowly, making sure that the hand spraying the paint does not get ahead of or lag behind the one holding the paint block. If you doubt your paint block abilities, stay on the safe side and mask with quality masking tape and paper or at least two sheets of newspaper sandwiched together.

Black semigloss paint generally fares well for visible frame members. If you are a stickler for detail, consider purchasing paint made for auto frames from The Eastwood Company or an autobody paint and supply outlet.

It is of utmost importance that the frame be clean, dry, and free of contaminants. Paint adheres best to clean, dry surfaces. Apply paint in a smooth and uniform fashion to avoid runs. Plan to apply two or three light coats, as opposed to a single heavy one. Hold the paint can about 6 to 8 inches away from the surface and overlap passes by 1 or 2 inches.

After the second coat, walk away from the vehicle and look at it from a distance. Squat down so your eyes will be at a level to see just how well you covered the visible frame. Touch up missed spots and continue the exercise until you are completely satisfied with the results.

Tailpipes

Exhaust tips and mufflers are best cleaned with a wet scouring pad. Use heat-resistant paint for these items. Choose a color that most fits your wishes, generally black or bright silver, that will help those parts best blend in with the rest of the vehicle. Exposed mufflers, like those on Corvettes and Jaguar roadsters, are plainly visible. Anything you can do to help them better blend in

Frame members can be painted any color you want. However, if you own a classic, you should stick with originality. Painting the frame underneath a vehicle is generally saved for show cars. You should focus your concern on those frame members that are clearly visible to other drivers and admirers.

with the bottom of the vehicle is an improvement.

The insides of tailpipes may be cleaned with rags and a scouring pad. Do not use much water for rinsing. If you think water has entered the tailpipe, simply start the engine to blow it out. The inside of the pipe can be painted a flat black, using heat-resistant paint.

Polish chrome exhaust tips with a chrome or all-metal polish. Stubborn spots may need the strength of number 0000 steel wool and polish. For bare end exhaust tips that stand out, consider painting them black or bright silver.

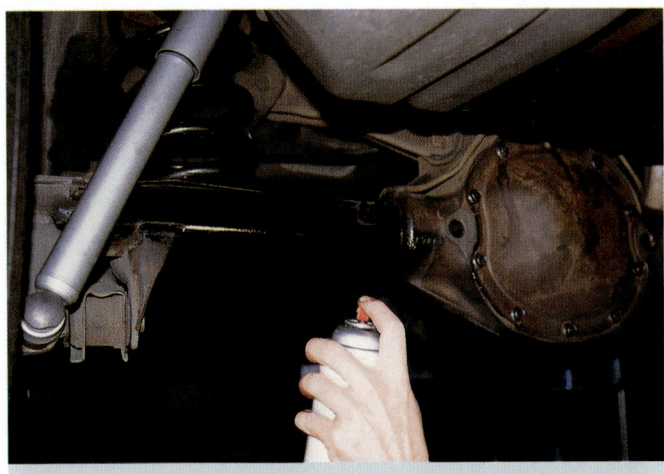

Semigloss black is commonly used to make rear axles look cleaner and newer. Visible shock absorbers should also be cleaned and painted, either a stock color or one of your choosing, as should fuel tanks and other plainly visible assemblies.

Differential covers have been painted different colors by vehicle owners, and some even switch to chrome-plated covers. The choice is up to you and how you want your ride to look.

Tailpipes are not generally thought of as much of anything, except to car people who appreciate the completeness of auto details. Cleaning and polishing tailpipes is another small step that helps to make one detailed vehicle look much nicer than one that has undergone a less intensive detail.

The tailpipe is looking much better now that it is being cleaned with number 0000 steel wool and polish. It will now blend in with the rest of the car to help the entire unit look great.

Chapter 7
Glass, Trim & Moldings

Don't forget about the details that can make the difference between "nice car" and "wow!"

Detailing Techniques

Have you ever noticed how much better an automobile seems to perform when the windows are clean? More importantly, have you ever considered clean glass to be a safety factor? Operating a motor vehicle on a rainy night, in heavy traffic, straining to see through a dirty windshield blurred by the smears of worn wipers should convince you.

Trim pieces in good condition work to accent a vehicle's unique bodylines. Loose ones, and those in need of detailing, can do exactly the opposite, making a car look used, worn, and beat up. The same can be said for molding. Torn, misaligned, and dried-out window and door moldings will not enhance the good looks of your automobile.

Taking care of the little things separates the best detailers from mediocre ones. Glass, trim, and moldings are important parts of any vehicle, and you need to pay attention to their condition, cleanliness, and appearance.

Clean windows are as important to a quality detail as shiny paint. They are one of the toughest parts to get perfectly clean, too. Under artificial light, they may look fine, but out in sunlight you will be able to readily notice smears, smudges, and other imperfections.

A wide variety of glass cleaners are available at auto parts stores and supermarkets. Some come as sprays and others are foam-based. Try a few of them until you find the one that works best for you.

TECHNIQUE 36 REMOVING STICKERS AND DECALS FROM GLASS

Time: 1/4 to 1 hour

Tools: Adhesive remover, razor scraper, clean cloth

Talent: ★

Tab: $5

Tip: Work slowly with a razor scraper to avoid slipping off stickers and onto something else

Gain: Improved vehicle appearance

Complementary project: Clean all window glass

Unsightly stickers and decals should be removed before glass cleaning is started, because you may have to use adhesive removers to loosen and dissolve glue residue. The sometimes-messy process can smear clean windows, requiring them to be washed again.

Stickers and decals are easier to remove from glass than from paint, because glass is more resistant to scratching. Try to peel them off with your fingernail. If that doesn't work, moisten a cloth with adhesive remover and soak the sticker to loosen the glue. Peel off as much as you can. Use the remover to take off the remaining glue.

As a last resort for stubborn stickers, carefully use a razor scraper. Don't try to remove the entire sticker in one swipe. Gently ease the blade into one corner of the sticker and slowly turn it a little at a time. Use your wrist to apply pressure, and not your arm. Controlled pressure is necessary to prevent the blade from slipping off the sticker and cutting into paint or upholstery.

Stickers on chrome bumpers are taken off using the same method. In lieu of a razor scraper, try using a hair blow dryer to loosen glue. Be careful, though—the sticker and surrounding metal can become quite hot.

Stickers and vinyl tape stripes on paint can be removed with an adhesive remover or hair blow dryer, but you risk the chance of paint underneath being a different color than the rest of the vehicle. Regardless of how long the sticker or vinyl stripes have been in place, the rest of the paint job will have suffered at least some oxidation. That exposed paint will likely be a hint lighter in color than the paint under the sticker or stripes. This is because the covered paint under the sticker has not been exposed to the sun as long as the rest of the paint. You can try to polish the area as a means to better blend the paint colors, but chances are the paint job will be left with a marred spot where the sticker or stripes were located.

You must be patient while removing vinyl stripes from painted surfaces. The glue loosens somewhat slowly and you have to be careful not to heat the painted surface too much. Overheating will cause paint to bubble. It is best to rely on an adhesive remover. Autobody paint and supply stores and most auto parts houses carry an assortment of sticker and decal removers, as well as other adhesive-removing products.

Below Left: An easy way to remove stickers from glass is to soften them up with heat from a hair blow dryer. Use an adhesive remover to get rid of any lingering adhesive.

Below Right: When using a razor scraper to remove stickers from glass, go slowly. The last thing you want to happen is to have the razor slip off the sticker and onto something nearby. Start at a corner and work the razor blade in an arc to loosen the adhesive's grip.

TECHNIQUE 37 GLASS CLEANING

Time: 1/2 to 1 hour

Tools: Glass cleaner, clean cloths

Talent: ★

Tab: $5

Tip: Pull vehicle out into sunlight for final inspection

Gain: Clean windows make everything in a vehicle feel clean

Complementary project: Apply Rain-X or similar water-shedding product to glass

As in choosing car wash soap, cleaners, polish, and wax, no two car enthusiasts quite seem to agree which window cleaners are best. Some prefer glass cleaners sold by dealerships, while others like run-of-the-mill products sold at supermarkets. A few auto buffs have even had good results using TSP (tri-sodium phosphate) or plain clean water. Owners of some show cars clean windows with a wash mitt rinsed in plain water and then polish glass with a household window cleaner. Others use either a regular glass cleaner or plain water and then dry glass with newspaper. They believe that the newspaper itself absorbs moisture, while the very, very fine grit in the ink helps to polish glass. As with other cosmetic car care products, you will have to experiment yourself until you find the combination that works best for you.

Glass cleaners in spray form work fine. The only problem may be overspray. It is difficult, at best, to direct these cleaners on the inside of a windshield without spotting the dashboard. You may have better luck spraying one side of a clean towel, wiping the glass with it, and then drying with the other side of the towel. Most glass cleaners, like Windex, are fine. Those containing ammonia tend to streak less. Sometimes, however, they tend to push a dirty film around windows to cause smears. Two or three applications may be necessary.

Liquid soaps will clean, but you must go over glass with clear water to remove the film they generally leave behind. Consider using liquid soap for only the dirtiest of windows. Keep in mind that glass cleaners will do the same job and will not leave that same kind of soapy film behind.

Although TSP is a good general cleaner, it is not widely considered a glass cleaner. As such, you should be concerned about tiny grains of TSP not fully dissolving in pails of water. Picked up by a washcloth or mitt, they become a scratch hazard. Considering that hazard and the extra steps involved with mixing, the solution may not be worth the effort.

Clear water with a clean wash mitt is simple and easy. There is not much worry about residual films or smears. A couple of swipes with the mitt, followed by a dry towel, should be sufficient in most cases.

Many detailers mix ammonia with water for glass cleaning. The ammonia

Cleaning windows does not require any special skill. Spray window cleaner onto a clean cloth and wipe. Refrain from spraying glass cleaner onto windows directly, as cleaner overspray may land on parts you have detailed.

Above Left: Be sure to clean to the uppermost parts of roll-down windows. Roll windows partway down first to clean both sides of the top section that fits into the molding recess. Afterward, roll the window back up to clean the rest of it.

Above Right: Cleaning windshields is easiest from the passenger side of the front seat. Cleaning rear windows is easiest from the rear seat on either side, depending on whether you are left- or right-handed. It may be most comfortable to use the back of your hand to clean rear windows. Take your time and have patience.

cuts through grease and helps the mixture to evaporate quickly. Not much is needed—about a capful per half-bucket of water.

Regardless of what glass cleaner you use, prevent the solution from dripping on painted surfaces or upholstery. If need be, lay a clean towel over the paint, dashboard, or seat for protection.

Start glass cleaning at the driver's door. Roll the window down an inch or so. This allows access to the top part of glass hidden in the upper groove of the frame. Clean and dry this part first. Then roll the window back up and clean the rest of it. Take note of the corners. This is where most of the smears will end up. Take your time and inspect each window before moving on. Don't forget the mirrors.

Rear windows are the most difficult to clean. Use the back of your hand to guide the towel or mitt. If necessary, unfold the towel and use a corner of it for more manageability. Be gentle while cleaning rear windows equipped with defrosters. The thin lines that run across the glass are actually small wires that heat up when they are activated to clear glass of moisture. A break in any of the lines will render the defroster useless.

The inside windshield is easiest to clean from the passenger side. The steering wheel won't be in your way and you can easily reach across the full length of the window. Go slowly around the rearview mirror glued to the windshield. Bumping into it with your hand at just the right angle will knock the bracket loose from the glass. Should this happen, use an adhesive specifically made for windshield-mounted mirrors. Other glue products just will not work nearly as well.

Cleaning Window Tint Film

Window tint film is common on many vehicles. A thin sheet of Mylar applied on the interior side of glass prevents folks from seeing inside to some degree, and helps to darken the effects of bright sun. Wash these windows with a mild window cleaner and water. Deep scratches cannot be repaired, as an entire new sheet will have to be installed.

Film-tinted windows that are extra dirty are cleaned with a nonabrasive cleaner designed for plastic windows, like those on convertibles. Meguiar's, Eagle 1, and other car care product manufacturers offer these products in both cleaner and polish formulas. Be sure to read the labels to ensure they are safe for plastic. Look for those that come in a spray bottle and are labeled safe for plastic, Plexiglass, and window tint film.

In addition to Mylar films, windows are also tinted with sprays. Again, you must clean these windows with care to avoid scratches. Use spray or liquid glass cleaners and soft, clean cotton cloths or towels.

TECHNIQUE 38 GLASS POLISHING

Time: 1/2 to 1-1/2 hours

Tools: Glass polish, clean cloth, Pro Glass Polishing Kit from The Eastwood Company

Talent: ★★

Tab: $10–$35

Tip: Start with the mildest methods first

Gain: New-looking glass without the cost for replacement

Complementary project: Wash entire vehicle

Some car folks wash windows with clear water and dry immediately with newspaper. The paper absorbs water and picks up dirt, while the ink works as a very fine polishing compound. Be careful around light-colored cloth interiors, as ink can rub off the newspaper to stain cloth door panels and seats.

Most cosmetic car care manufacturers offer glass cleaner products. They are designed to dissolve road grime, dirt, tar, and oil, and to remove films caused by cigarette smoke and vinyl vapors. Apply glass cleaner like wax; pour or spray a small amount on a damp sponge or cloth and wipe on a thin coat in a straight back-and-forth motion. Let it dry to a haze and remove with a clean soft cloth. These products normally wipe off without a dusty residue.

Paint overspray and dried insect residue can be removed from glass in a few different ways. First, always try water and a soft cloth. For more persistent residue, like tar and paint overspray, try a glass cleaner/polish product. Auto parts stores usually carry a number of brands. If glass polish fails to remove stubborn residue, try number 0000 steel wool and glass cleaner together. Make sure you are working on clean, bare glass and not something with a tint film over it, as steel wool will surely scratch such films. Afterward, clean the glass again with glass cleaner and a clean cloth to remove any remaining steel wool fibers. Note: Steel wool may scratch acrylic windows on newer vehicles. For those, use a plastic cleaner followed by plastic polish. Meguiar's offers both. If you are not sure whether the windows on your newer automobile are glass or acrylic, look closely for a small label at one of the lower corners of the windows. If the label doesn't help, ask the service manager at a local dealership or auto glass shop.

With the failure of these processes and the thought of glass replacement, try buffing glass with a regular paint buffer and mild sealer/glaze. In some cases, this works to remove etching and severe water spots.

The Pro Glass Polishing Kit from The Eastwood Company is designed to remove light scratches, haze, and wiper marks from windshields. It will not polish out deep scratches that you can catch with a fingernail. Be sure to follow instructions carefully to ensure a good job. The slurry is mixed with water; the felt wheel must be wet and not run over 1,500 rpm.

Glass, Trim & Moldings 115

TECHNIQUE 39 REMOVING BUILDUP FROM WINDOW EDGES

Time: 1/2 hour

Tools: Glass and all-purpose cleaners, toothbrush, towels, cotton swabs, razor scraper

Talent: ★

Tab: $5

Tip: Once cleaned, go over these areas with a paintbrush while washing the vehicle to prevent future buildup

Gain: Crisper-looking windows all around

Complementary project: Dress moldings with a light coat of vinyl/rubber protectant

A close inspection may reveal buildup of dirt in the corners of glass, next to trim. This is especially prevalent along windshields, rear windows, and fixed glass. Remove this caked-on stuff with a toothbrush and cotton swab. Lay a towel down at the base to catch drips and splatter.

Spray a toothbrush with window cleaner, or dip it in water and shake off the excess. Scrub along the trim and molding to break the crud loose. Use the fold of a thin towel to reach into seams and remove debris. Use a cotton swab in tighter spots. If a toothbrush isn't strong enough, gently slide the blade of a razor scraper along the glass and under the dirt. Carefully raise the back of the scraper to loosen the dirt. Follow with a toothbrush and towel.

Special care should be taken when using a razor scraper. Inserting it too far into molding may cause leaks; a slip could end in a scratch or tear; and while pushing it along an inside window curve, the square edges could scratch glass. Use this tool as a last resort.

Defogger strips on rear windows are not imbedded in the glass. For windows with strips on the surface, wipe in the direction of the strips, not against them. Never use a razor scraper near them. This same caution applies to areas on the molding where defogger strips exit from the glass. The slightest break in this continuous strip will render the defogger useless.

Vehicles that have been parked outdoors for long periods of time may suffer mold or mildew buildup around outer window moldings. Look closely for any signs of dark spotting or green-colored buildup. Problems of this nature are most common in areas with rather wet climates. They can also persist on vehicles regularly parked under trees.

Remove mold and mildew accumulations with cleaner and a toothbrush. Afterward, wipe off areas with a clean damp cloth. If accumulations are significant, you may have to resort to a stronger cleaner. There are times when you may have to use a scrub brush with a powdered cleanser that contains a bleach additive. In these cases, be certain to cover the area of the vehicle below the windows with towels or plastic. Most powdered cleansers with bleach are potent products, and you don't want a slurry of the cleaner to come in contact with paint, trim, emblems, badges, special wheels, and so on.

Areas with lots of trees, frequent rain, and high humidity will often cause vehicles to accumulate a buildup of dirt and crud along window molding edges. Use a toothbrush or soft scrub brush to get window trim edges clean.

TECHNIQUE 40 TRIM CLEANING

Time: 1 hour

Tools: Soap and water, paintbrush, toothbrush, bug and tar remover, clean cloths

Talent: ★

Tab: $5

Tip: Use a toothbrush to get dried polish and wax out of emblems

Gain: Crisper-looking trim all around

Complementary project: Apply a light coat of vinyl/rubber dressing to vinyl and rubber trim

Trim includes metal strips surrounding glass, headlights, taillights, and vinyl tops, as well as the straight strips that run the length of some automobiles. They are attached by means of screws, clips, adhesive, and even double-backed tape strips. Regardless of the means of attachment, dirt will find its way into, behind and along trim edges.

The easiest way to clean most trim is to use a paintbrush during the preliminary wash. The skinny bristles reach into grooves on the face and seams to break loose dirt, pine needles, and polish/wax residue. Gently use a toothbrush to clean screw slots and stubborn wax and dirt buildup along seams and edges.

Direct a small stream of water behind the body side trim to dislodge and float away dirt caught between it and the car body. Lower rocker panel trim may have to be cleaned with bug and tar remover to get rid of road tar blemishes.

Bona fide chrome trim pieces are shined with chrome polish. Alloys and softer metals must be polished with a product designed for such, like mag or all-metal polish. Using an abrasive polish on soft metal components will result in scratching. Many times you will have good results using mild auto polish like Meguiar's Number 7 for trim pieces. Start with the mildest polishes first before graduating to stronger formulas. Remove dried polish with a soft cloth and a 1-inch-wide paintbrush with the bristles cut to about 1/2 to 3/4 of an inch long. The brush works great for dislodging dried polish and wax from tight spaces on and around trim.

If you find yourself with a weathered classic sporting true chrome trim pieces in need of aggressive polishing, take time to mask painted areas around the trim pieces with masking tape. Aggressive polishing on true chrome pieces may entail polish and the use of number 0000 steel wool to rub out bad spots and bring back the deep shine chrome is so well noted for. Masking off painted areas around chrome will protect paint from mishaps, should you misguide a polishing swipe and your hand slides off the trim. Use the same precaution around glass, too.

Some trim is accented with painted stripes. If the paint is chipped or worn, repaint as necessary. Depending on the paint flaws, you may be able to complete small paint chip repairs with an artist's paintbrush. Larger flaws might require light sanding, good masking, and a spray paint endeavor. Check with an autobody paint and supply store for matching colors.

A steady hand helps to apply paint in even strokes. If the piece you are repainting is longer than you feel comfortable free-handing, purchase a roll of thin blue masking tape at the autobody paint and supply store

Trim, light housings, body side moldings, and the like are easiest to clean with a soft cloth and a floppy paintbrush. Agitate cleaner with the paintbrush to loosen dirt and grime buildup. Wipe away residue with a clean cloth.

Glass, Trim & Moldings **117**

Stubborn dirt, dry polish, and wax are quickly removed with a cut-off paintbrush. A 2-inch-wide paintbrush with the bristles cut to about 1/2 to 3/4 of an inch works well. Place a piece of tape over the brush's metal band to prevent paint scratch mishaps.

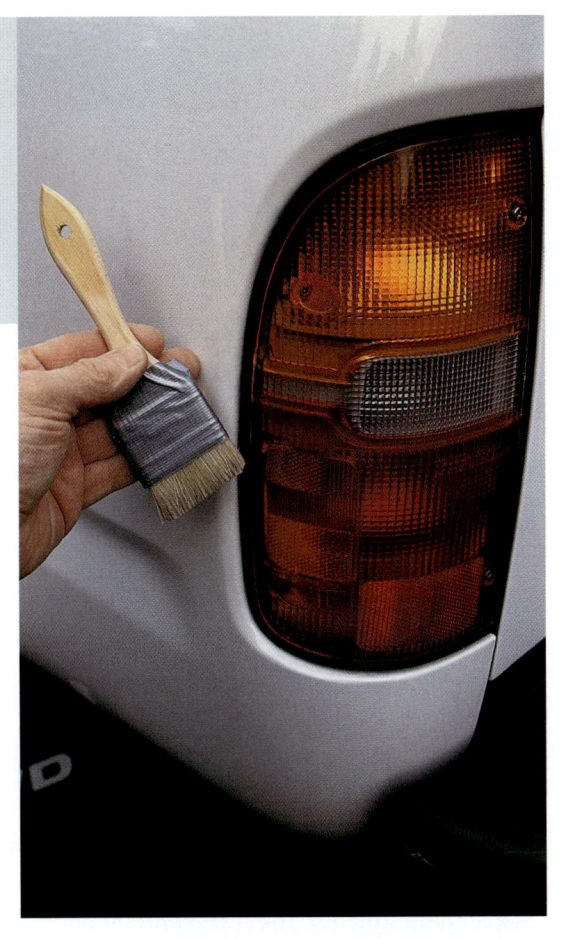

or at a sign painter's supply house. Ask for the kind of tape that pinstripers use. It seals tight, is easy to maneuver, and can be stuck on and peeled off a few times before losing its adhesiveness, a blessing when attempting to place it perfectly straight.

When you are done using the fine artist's paintbrush to apply paint, clean it thoroughly and apply a thin coat of petroleum jelly to the bristles. This helps them to retain their shape and prevents them from drying out into a frizz ball. These paintbrushes are available at hobby shops, artist's supply stores, and sign painter's supply houses.

If your car is equipped with plastic trim on the interior and the pseudochrome plastic film over the trim is peeling away or falling off, try this. Peel off all of the film. Sand the trim smooth and paint it glossy black. You will lose the chrome appearance but save the cost of new trim.

Black Trim

Many newer cars feature black trim. Most of this newer trim is plastic, although a lot of pieces are made of rubber, vinyl, or anodized aluminum.

Plastic trim is cleaned with a wash mitt, paintbrush, and toothbrush using mild soap or liquid cleaner. It can also be polished with either Meguiar's or Mothers plastic polish. Solid textured, smooth plastic can be waxed along with the rest of the car. Textured plastic cannot be waxed, because wax will adhere to the grain and be very difficult to remove. Reportedly, a little light scrubbing with peanut butter can remove ugly wax buildup from such textured surfaces. Lightly dress those parts with an all-purpose rubber and vinyl protectant. Rubber trim is cleaned and dressed the same way. If you are in doubt whether to wax or dress, test a small, inconspicuous spot first to determine product compatibility.

Anodized aluminum is fragile. The sun is its worst enemy, as it beats down and fades the black color. Harsh rubbing and scrubbing will also quickly remove the thin-coated surface. Wash with a clean soft mitt and a paintbrush, using mild car wash soap. Once or twice a year, apply a very light coat of wax. Never use polish, as the slightest grit will surely blemish or remove the anodization. If anodization is eventually removed, you'll have to either have the part reanodized or sanded down and painted.

Meguiar's Trim Detailer is designed to keep plastic trim and rubber moldings looking new. Use it to bring back the luster and color to faded plastic bumper trim, moldings, window trim, and the like. Apply with a clean soft cotton towel to cool surfaces. Buff off excess with a dry part of the towel.

TECHNIQUE 41 CHROME, RUBBER, AND VINYL

Time: 1/2 to 1 hour

Tools: Chrome or all metal polish, vinyl/rubber protectant, clean cloths

Talent: ★

Tab: $5–$15

Tip: Go slowly to avoid smearing paint near trim pieces

Gain: Brighter vehicle accents

Complementary project: Clean and dress tires

All chrome, including bumpers, mirrors, door handles, fender moldings, hood ornaments, and trim, should be polished and waxed. Happich Simichrome works very well, as do most chrome polishes. As with paint, don't polish just for the sake of polishing. Do it when needed and do it gently.

Note: Not all bright metal parts on automobile exteriors are real chrome! They were on older cars, but not on the newer ones of the last few decades. Using a metal polish not designed for all metals may result in unsightly scratching. You must be cautious! Read the label on the polish product you intend to purchase to ensure it is safe for the types of metal you intend to brighten.

Heavily rusted and pitted chrome may be cleaned with a wet scouring pad or number 0000 steel wool and polish. Follow this with a mild chrome polish. Afterward, apply a thin coat of carnauba wax. Use a soft cloth to remove polish and wax. A cut-off paintbrush is great for removing buildup around carriage bolts, screw slots, and the like.

Badly pitted chrome parts must be rechromed at a plating shop; they just can't be properly repaired in any other fashion. However, to make those parts look decent during the interim, clean and buff with number 0000 steel wool and polish. Before applying wax, clean the parts again with Prepsol or another wax-removing solvent. Then, using a toothpick, the clean end of a cardboard match, or a fine artist's paintbrush, dab a bit of bright silver paint into the exposed pits and other areas where

Below Left: Place a dab of chrome polish on a clean soft cotton cloth and wipe on. You should begin to see the pleasing results right away. Turn the cloth over to a clean side to wipe off residue. Place your fingers inside the cloth and use them as guides to ensure you do not rub chrome polish onto any other surfaces.

Below Right: Older classic automobiles are loaded with chrome. Some time and patience were needed to get all of the chrome on this vintage car in pristine condition. Careful application of chrome polish prevents accidental smears on paint, especially while polishing body side and window trim.

Glass, Trim & Moldings 119

Do not spray vinyl/rubber protectant directly onto trim or molding. It is best to spray some onto a clean cloth away from the vehicle. This prevents unwanted overspray on parts already polished and waxed.

Rub a cloth dampened with vinyl/rubber protectant onto trim and moldings to make them look new. Be sure to fold the cloth in such a way that it does not drag onto painted surfaces. If the cloth you intend to use is large, cut it down to a manageable size. For this kind of application, a cloth about the size of a normal bathroom washcloth is about right.

chrome has peeled away. At a distance, the part will look fine, depending upon the size of the touch-ups.

Rubber and Vinyl

Severely neglected rubber or vinyl trim is rejuvenated with a little effort. First, scrub the trim clean, using a scrub brush and liquid cleaner. Be sure you have cleaned all of the dirt, mold, mildew, and other blemishes from the surface. Trim may look faded and a much lighter color than desired once it has dried.

Next, bathe the trim with a good vinyl and rubber conditioner like Meguiar's Number 40. Use a paintbrush or soft scrub brush to really work in the conditioner. Use plenty to completely saturate the material. Wipe off the excess from painted areas around the trim and then let it soak in for a few hours, even overnight.

The next day, or a few hours later, whip up a bucketful of car wash soap and wash all of the rubber or vinyl completely. Clean the painted areas around trim, too. Use a wash mitt, as there should be no need to scrub. Do a good job of washing off as much of the excess conditioner as you can. Then dry. The results should reflect vinyl trim that looks almost as good as new, with no smudged conditioner residue on the paint. A thorough washing will remove the excessive sheen and slippery texture on the material to leave it looking dark and rich.

Special vehicles often present detailers with unusual tasks. In this instance, the detailer is treating the wood on the running board of a pristine vintage classic Clenet. A high-quality wood polish, rather than a regular automotive protectant, has been applied and is being buffed off.

TECHNIQUE 42 MOLDINGS

Time: 1/2 hour

Tools: Soap and water, toothbrush, cloths, vinyl/rubber protectant

Talent: ★

Tab: $5

Tip: Be gentle on moldings to prevent tearing or breaking them loose from adhesive

Gain: Molding longevity

Complementary project: Clean doorjambs and frames

Rubber moldings are located around doors, trunk lids, and windows. These items should not be overlooked during a detail. Clean them with soap and water, and use a toothbrush with liquid cleaner on stubborn spots.

Paint overspray is removed with lacquer thinner, but be extremely careful. The slightest drop of lacquer thinner will blemish paint. Just put a dab on a rag away from the vehicle and wipe the molding. It is best to make a number of gentle passes instead of one heavy swipe. Gentle pressure is also advised, especially for the soft moldings along doors.

Pickup trucks display heavy rubber molding around rear windows. It is very common for these moldings to be laden with dirt. Use plenty of soap and water to clean, using a small brush as necessary to loosen caked-on buildup. Plan to clean dirty moldings like these more than once to get them looking their best.

After rubber moldings have had time to dry, rub on a thin coat of vinyl and rubber conditioner. This will improve their appearance and prolong elasticity. To prevent smearing the glass while applying the protectant, use a clean paint block. Place a thin piece of cardboard down on top of the glass and next to the molding to serve as a nice guard. Protectant dressing will touch the guard and not the glass.

The rubber window moldings around pickup truck rear windows are frequently caked with dirt and road grime. Use a soft scrub brush and cleaner to remove contaminants and get them looking new again. After moldings have dried, treat them to a light coat of vinyl/rubber protectant.

A paintbrush works well to reach inside recesses and other tight areas along moldings. Spray a little all-purpose cleaner on the bristles and agitate them along dirty molding sections. Wipe off residue with a soft cotton cloth. Once the surface is dry, apply a light coat of protectant to moldings.

Glass, Trim & Moldings   121

TECHNIQUE 43 WINDSHIELD REPAIR

Time: 1/2 hour

Tools: Windshield ding repair kit

Talent: ★★

Tab: $10

Tip: Make sure you follow instructions

Gain: Clear windshield

Complementary project: Clean all windows

Bull's eyes and other windshield dings can be repaired at a professional shop, like Novus, or you can tackle the job yourself. Most auto parts stores carry windshield repair kits that are easy to use and do a good job of hiding damage and preventing glass from cracking.

In essence, windshield repair is accomplished by forcing a resin into a blemish. Dings are basically small craters that have suffered the loss of material in and around them. The resin material that is forced into them fills in the spots of missing glass. The best jobs result in resin being forced into the tiny fingers that spread away from the crater to make the repair virtually invisible.

It is very important that windshield dings be repaired as soon as possible. Allowing them to linger will do nothing but intensify the problem. Vibration can cause cracks to spread out of the ding and across the glass. Moisture and contaminants can also enter the space, making it tougher for the resin to form a nearly invisible bond.

Some repair kits employ a plungerlike tool that is operated up and down to force resin into the ding. Others employ a different means, but generally speaking, you have to be careful to prevent excessive pressure from causing more damage. Be sure to read and follow instructions carefully.

If your car suffers a windshield ding while you are on the road, pull over and put a piece of clear tape over the damage. This will protect it from moisture and contaminants. Get it fixed as soon as you can.

If you are unsure about fixing a windshield ding yourself, call your insurance company. Many auto insurance agencies will cover the cost of a Novus repair with no cost to you.

Windshield repair kits sell for around $10 in auto parts stores. The simple process forces resin inside windshield dings to seal up openings and fill cracks. Repairs are almost invisible and a lot cheaper than a new windshield.

Windshield wipers in need of new paint are first cleaned and then sanded, if necessary. They can be painted in place as long as sufficient masking is done first. Use good masking paper or doubled newspaper pages. A small piece of wood holds the wiper off of the windshield. Remove the blades before sanding or painting.

TECHNIQUE 44 PLASTIC WINDOW CARE

Time: 1/2 to 1 hour

Tools: Plastic polish, clean soft cloth

Talent: ★

Tab: $10

Tip: Start with more than one attempt with the mildest method first before graduating to more aggressive methods

Gain: Ability to see through plastic windows

Complementary project: Clean the convertible top

Dusty plastic windows on convertibles are cleaned with lots of water and a very soft cloth filled with soapsuds. Meguiar's Plastic Polish Number 10 and Plastic Cleaner Number 17 are products designed specifically for plastic window cleaning and polishing. Do not use any cleaning or polishing product on plastic windows unless the label states that it is designed for such applications.

Plastic window material is very fragile, and it scratches easily. Using a product with abrasives in it will cause scratches, some to the point that windows will be clouded and you will not be able to see through them. Of equal concern is the cloth you use to apply cleaner and polish. Make sure it is clean and very soft, like an old flannel shirt, a baby diaper, or a soft cotton towel. Use one cloth for cleaner and another for polish.

Plastic window cleaners are initially used to remove the big stuff. Polishes are then used to shine plastic windows to perfection. These products fill scratches and leave a protective film over the windows. This film helps to reduce water spotting and dust accumulation. These products are also good for Plexiglas and window tint film. Severe scratches will require several applications.

The Eastwood Company offers a vinyl window polish kit. It requires the use of an 1,800- to 3,500-rpm drill, polisher, or buffer. The kit sells for around $20. The deluxe kit with pads sells for around $30.

Use the right materials to detail plastic windows. Auto parts stores carry an assortment. Meguiar's Clear Plastic Cleaner will remove fine scratches and other blemishes. It can be applied by hand or with an orbital buffer. Clear Plastic Polish takes it to the next step to make plastic windows as clear as they can be. Use Clear Plastic Detailer on a regular basis to keep plastic windows looking their best.

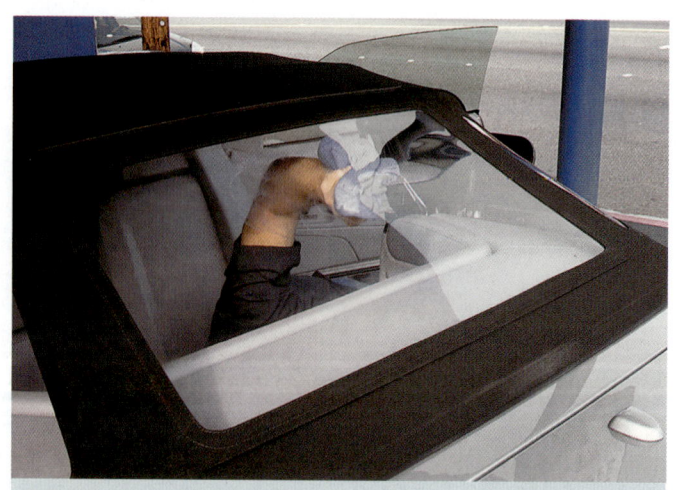

There is no need to rub hard on plastic window surfaces. Let the cleaners and polishes do the work. Remember to buff off residue with a separate clean soft cloth.

Glass, Trim & Moldings 123

TECHNIQUE 45 ALL THE LITTLE EXTRAS

Time: 1/2 to 2 hours

Tools: Cleaner, toothbrush, cut-off paintbrush, soft clean cloths

Talent: ★

Tab: $5–$25

Tip: Inspect vehicle in sunlight for optimum clarity

Gain: Detailing perfection

Complementary project: Dust off the entire vehicle

ID Badges

ID badges, such as the Porsche or Lamborghini crests, are cleaned with a soft toothbrush and mild soap. Those suffering exceptional buildup may be gently scrubbed with a toothbrush and mild polish; use very gentle pressure so you don't polish away the thin film of paint on the various adornments. Use a carnauba-based wax for protection and a very soft cloth for buffing. Wax residue along tiny ridges is removed with a soft cut-off paintbrush.

If you have to touch up paint on badges or crests, use a fine artist's paintbrush. They are available at hobby shops, artist's supply stores, and sign painter's supply houses. Buy paint at an autobody paint and supply store. Often you'll find that the store will carry a small assortment of paint in vials for such intricate work.

If you need to complete some paintwork around the outer edge of badges, you must take pains to do a good job of masking. Clean and wax the badge first so that masking tape will not adhere too strongly to the badge. This will prevent paint from being peeled off when the masking tape is removed. Carefully apply tape to the badge, overlapping strips to ensure full coverage. When the badge is completely covered, use a sharp knife, like a razor-bladed X-Acto knife, to trim the tape along the badge base edges. Use the knife at a slight angle in order to cut the tape at the very base of the badge and prevent overspray on its side.

License Plates

License plates should be removed from vehicles at least once a year so you can clean both the front and back. Clean them with a mitt, paintbrush, and liquid cleaner as necessary.

Should the plate on your car be oxidized, gently shine it with a mild polish. Exces-

Light work with a toothbrush and cotton swabs have made this piece look good. Dust has settled around it because the car just finished a slalom at a local concours event. Bringing it back up to its first-class appearance will be easy.

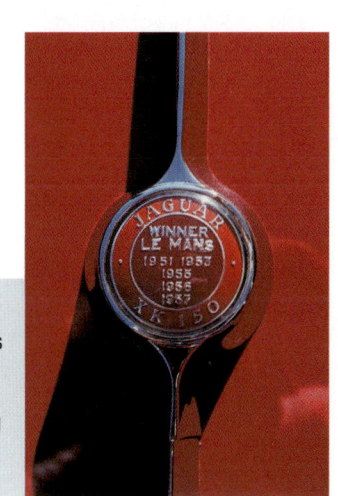

License plates and their frames are eye catchers when they are crooked or are damaged. Make time to ensure the license plates on your vehicle feature the kind of frames you want, that screws are not rusted, and that tabs are put on straight.

124 *Detailing Techniques*

Far Left: This rear taillight looks most impressive, thanks to a cut-off paintbrush and plastic polish. Notice that it is not too glossy but exhibits just the right luster. The chrome trim is polished to perfection, and there is no hint of dry polish or wax anywhere.

Left: The little extras include polishing such things as chrome spotlights. Nothing must be overlooked. Each section of the automobile and every part in those sections must be cleaned and polished. Make good use of the cut-off paintbrush to ensure that grooves, slots, and recesses are free from unsightly debris.

sive polishing could remove paint from the letters and numbers, so go easy. The plate may also be waxed for protection.

When it comes time to install new tabs, take the license plate off the vehicle and bring it into the house. Wash it in the sink and towel dry. Allow the new tabs and the license plate to warm to room temperature. This makes application of the plastic tabs easier and ensures their best adherence. Make sure you put on the month and year tabs straight and even.

Be certain you like the license plate holder on your automobile before putting the plate back on. If you do, make sure it is clean and waxed appropriately. Use the right kind of screws to attach the plate and holder to the vehicle, and make sure the screws match. If you do not particularly care for the license plate frame, discard it. Or, make a run down to the auto parts store and get a new one. They cost anywhere from $5 to $15.

Plastic Light Covers

Scratches on plastic light covers are polished with plastic cleaner and polish. Carefully do this with the part on the car, or better yet, take it off and do it away from the vehicle. This way you can clean the inside, too. Be sure to use a cut-off paintbrush to remove all traces of polish and wax residue.

Cracked plastic light covers can be repaired. The Eastwood Company and most autobody paint and supply stores sell kits designed specifically for plastic lens repair. These kits work well for repairing minor cracks, chips, and holes. They include colored translucent adhesive epoxy compounds in amber and red that actually replace missing lens material.

Antennas

Radio and telephone antennas should not be overlooked. Those in moderately clean condition are washed with a mitt at the same time the car is washed. Those with accumulations of dirt and grease are cleaned with a paper towel sprayed with WD-40 and followed with a quick buff by a clean cloth. Extra-dirty antennas are gently scrubbed with a wet scouring pad, but make sure you rinse away all pad residue from the antenna and car body. Clean antenna bases with a paintbrush or toothbrush.

Chrome antennas are polished with chrome polish and then lightly waxed. Electric antennas should be wiped with a cloth sprayed with WD-40 or similar lubricant. A light coat of lubrication helps power antennas to extend and retract smoothly, prolonging their operation.

Glass, Trim & Moldings 125

The only thing standing out on this vehicle is the tires; they need to be cleaned after this little outing. Notice that the chrome is polished and the license plate hangs square. The running boards are tidy, and the glass is clean. All parts, large and small, blend nicely together to make this a most pleasant vehicle to admire.

Windshield and Other Wipers

Windshield and headlight wipers work best when rubber is in good shape and the glass is clean. Wash the front and rear wiper blades at the same time you wash the glass. Use a paintbrush and a toothbrush to remove bug residue, polish, wax, and any dirt or grease buildup. The toothbrush works well to clean rubber and the metal parts used to make up the entire assembly.

Although dressing helps most other rubber parts maintain a rich dark color, wiper blades should not be treated. Dressing will quickly wash off and under light misty conditions could cause smears on glass. The best overall treatment is frequent cleaning.

No amount of cleaning will make wiper blades last forever. When they become faulty, worn, or dried, replace them with a new set. A once-a-year replacement is normal for windshield wipers, rear window wipers, and those on headlights.

The metal parts on wiper assemblies are cleaned with soap and water and a paintbrush or toothbrush. Those that are designed to shine are polished with a mild product designed for all types of metals, like Meguiar's All Metal Polish. Using some chrome polishes on softer metals can result in scratches or general surface marring. Be sure to read the labels of products you plan to use to ensure they are compatible with the metal on your vehicle's wipers.

Black wiper assemblies can be repainted as necessary. It may be best to pull them off of the vehicle and paint them separately. If they just need a slight touch-up, you can leave them on the car. Use masking paper or double-thick newspaper to mask between the blade and the vehicle's cowling, hood, and windshield. Make sure the rubber wiper is removed before painting. You can use a can of spray paint or dab paint from a can with a fine artist's paintbrush. Depending on what kind of paint came on the wiper, use either flat or semigloss black or bright silver.

As a means to help shed rainwater and keep interior glass surfaces from fogging up, consider applications of Rain-X to both sides of the windows. Two formulas are available, one for the exterior and one for the interior.

Chapter 8
Tires & Wheels

Wheels carry tires. Tires ride on the ground. The ground is dirty. You do the math.

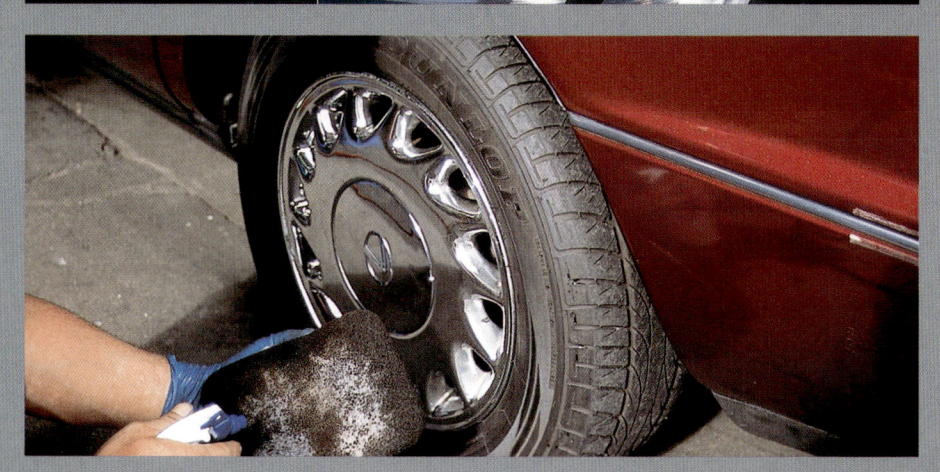

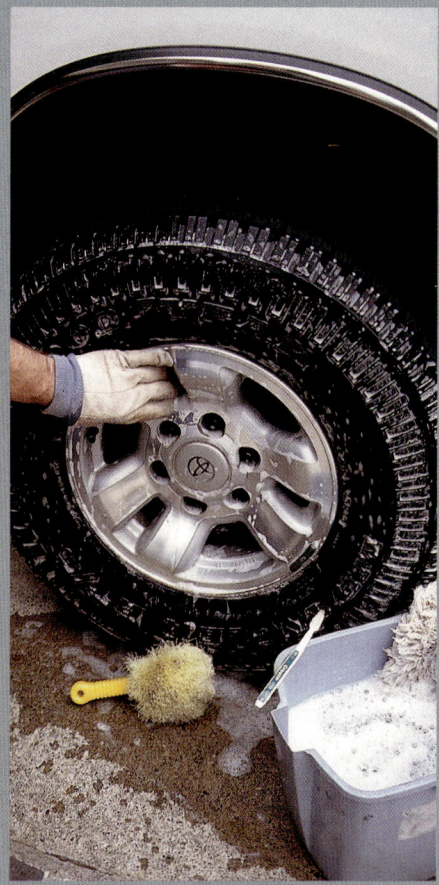

Tires and wheels have come a long way in the last few decades. Car people treat wheels and tires as part of their automobile's overall cosmetic appearance, rather than as a mere necessity. Frequently, the right wheels and tires are the final touch to really make a car stand out, especially when they are meticulously maintained.

Along with the new breed of wheels comes the dilemma of cleaning. You can use an all-purpose wheel cleaner with some results, but what is it actually doing to the surface of the wheel? Is it all right to use chrome cleaner on mag wheels? Will whitewall cleaner blemish certain wheels? If so, how do you get the whitewalls white? Plan some time to clean wheels and tires, including the backs and the spare.

Do the wheels and tires on your car look nice from a distance but just so-so up close? Next time you admire an exhibit at a car show, take note of the wheels and tires. You will see that they are cleaned and polished inside, as well as outside. Most new wheels are slotted, allowing the backs to show through. Why spend an hour cleaning the front when dirt is still visible from the back?

Choose a suitable work place and have cleaners, brushes, and a garden hose handy. Use a sturdy jack to raise the vehicle, and place jack stands under it. Then remove a wheel and tire. Clean, polish, and wax one wheel and tire at a time. Before replacing, consider cleaning and painting the fenderwell and visible underbody.

Detailed tires and wheels help make automobiles look crisp and stand tall. You wouldn't wear dirty shoes with new clothes, nor should you ignore tires and wheels when the rest of your car or truck looks new. Notice how the tires and wheels on this vintage classic add to its overall beauty.

This tire will clean up nicely with all-purpose cleaner and a scrub brush. A paintbrush will be most useful in helping to remove the stuff caked onto the slots along the wheel perimeter and the groove around the center cap.

Tires & Wheels   129

TECHNIQUE 46 WHITEWALLS AND RAISED WHITE LETTERING

Time: 1/2 to 1 hour

Tools: Soap and water, cleaner, scrub brush, whitewall wire brush, scouring pad

Talent: ★

Tab: $5

Tip: Rinse fenderwells completely before starting

Gain: Crisp-looking whitewalls and raised white letters

Complementary project: Dress tires and polish wheels

Use whitewall cleaners as you may, always following instructions on the label. Note label cautions such as, "Clean one tire at a time, and do not allow cleaner to dry on chrome, polished aluminum, or painted wheels." Chemicals in some cleaners are strong and may blemish certain wheel surfaces.

Instead of whitewall cleaners, you can use car wash soap or dish soap and a whitewall brush. This small wire brush features short bristles cropped closely together. Used with soap, it scrubs whitewalls and raised white letters to their brightest.

Some detailers prefer to use wet scouring pads. The combination of soap and soft steel wool makes short work of whitewall cleaning. The side of the pad fits into thin whitewalls and into tight grooves where the outer white ring butts against the black.

This is a whitewall cleaning brush. It features short wire bristles cropped closely together. It does an excellent job of cleaning up whitewalls and raised white letters when used along with an all-purpose cleaner.

1. After rinsing the tire with clear water, spray an all-purpose cleaner like Simple Green or Meguiar's Extra all over the tire surface. Use a plastic-bristled brush to scrub the whole tire. Follow up with the whitewall brush on whitewalls or raised white letters. Rinse with water and repeat as needed until the tire and whitewall are sparkling clean.

Detailing Techniques

For tough scuff marks, try a scouring pad and Simple Green. Rinse with plenty of water after each application. Keep at it until you are satisfied with the results. Minor divots and lightly damaged whitewalls can be touched up with quick-drying, flat white latex paint. Major damage must be repaired professionally. Check with local detail shops and auto dealerships to find out who does this kind of work in your local area.

2. Some detailers rely on scouring pads, rather than a whitewall brush, to help clean and brighten whitewalls and raised white letters. Many prefer steel wool scouring pads like SOS, while others have enjoyed good results using Scotch-Brite pads. Be sure to use plenty of liquid cleaner along with the pads, as it serves as a lubricant along with providing added cleaning power.

3. Use a clean towel to dry tires after they have been cleaned and rinsed with water. This will help you recognize missed spots and also prepare the surface for a coat of protectant dressing. Dirty smudges should be touched up with the whitewall brush or scouring pad.

4. Spray a coat of tire dressing onto the tire and buff it in with a soft brush or cloth. A brush works well to get dressing into the slots along the perimeter of the tires and around the seams surrounding lettering.

5. Once the tire has been completely coated with dressing, use a dry cloth to buff off all excess. This will help to remove the excessive sheen left on the surface to make the tire look crisp and new. Use a dry steel wool scouring pad to dress up whitewalls and raised white letters and to remove the light blue tint left on them by the dressing.

Tires & Wheels

TECHNIQUE 47 DRESSING AND TIRE BLACK

Time: 1/2 hour

Tools: Tire dressing (vinyl/rubber protectant), soft brush, clean cloth

Talent: ★

Tab: $5–$10

Tip: Be sure tires are clean and dry before application

Gain: New-looking tires

Complementary project: Clean and paint fenderwells

There are a few auto enthusiasts who believe tire dressings can cause tires to split and crack after repeated weekly applications. They think that the solvents used to open the pores on tires to allow silicone to sink in might cause rubber to dry out and crack or split. One could surmise that excessive dressing applications with a failure to wipe off the excess could potentially result in such tire damage. So remember to follow label instructions and wipe off excess applications.

Other auto enthusiasts have used tire dressing regularly for years. They apply as needed and buff tires off with a clean cloth afterward. Sometimes they simply use the regular dressing cloth by itself to touch up minor scuffs, as it will be saturated with dressing from previous applications.

It is better to apply a tire dressing like Meguiar's Vinyl & Rubber Conditioner Number 40 or Mothers Duration Extended Wear Tire Care than tire black. Tire black has its purpose and does a good job of making ugly tires look better, but an intensive cleaning and proper application of conditioner dressing make most tires look clean and original.

Regular tire dressings like Meguiar's Number 40 and Armor All can be sprayed or wiped on. Spraying will likely result in some overspray onto the wheels but will reach deep into crevices on sidewall tread and lettering. A good compromise might be controlled spraying of the outer edge followed by buffing with a cloth or soft brush to work in the dressing.

For a heavy-duty one-time tire treatment, soak the tire with lots of dressing and let it stand for a few hours. Then wash with mild soap and water and dry. The tire should look brand-new without the glossy look of being freshly dressed.

Use the edge of a soft brush to spread dressing along the border between tire and wheel and into grooves and tread patterns. Follow with a clean towel to buff off excess. Go over whitewalls and raised white letters with a clean, dry scouring pad to remove the bluish tint from them.

Tire dressing can be sprayed directly onto tires and then buffed in, or it can be sprayed onto an applicator and then applied to the tire to avoid overspray on the wheel. Expect to apply more dressing onto the applicator to ensure thorough coverage.

Detailing Techniques

This detailer is applying a liberal amount of dressing onto the tire. The applicator is moved around in different directions to ensure adequate coverage.

The tire is covered with a heavy coat of dressing that will be left in place for a while. After the dressing has time to soak into the rubber, the tire and wheel will be thoroughly washed and cleaned. The wheel needs cleaning, of course, and the second tire cleaning will remove the excessive sheen from the tire to leave it looking new and crisp.

At the auto parts store, you will notice different types of tire dressings that are creams, as opposed to sprays, such as Meguiar's Endurance and Mothers Duration. These products come with their own foam applicator. There is enough creamy product in the bottle to keep four regular-sized tires looking good for a year. Make sure the tire is clean and dry, and then apply a liberal amount of Endurance or Duration to the applicator. Spread the gel evenly on the surface of the tire until a uniform appearance is achieved. A second coat may be applied for a shinier look.

Neither product includes solvents. Instead, they employ high molecular weight polymers and advanced film formers to coat tires with a protective shield. They are also designed with built-in UV inhibitors that allow rubber to breathe.

Spare tires should be cleaned and lightly dressed along with the other tires. Spare tires that sit idle in trunks for long periods tend to dull, and the dust that settles on them will accumulate to make the surface appear stained. Brighten whitewalls with a scouring pad.

Along with spray-on liquid tire dressings, you will also find creams and gels at auto parts stores. Mothers Duration was specifically designed to let tires breathe so the built-in UV inhibitors can do their job. The supplied applicator makes the dressing easy to apply without worry of overspray on the wheels.

TECHNIQUE 48 CHEMICAL WHEEL CLEANERS

Time: 1 hour

Tools: Correct chemical wheel cleaner, water, bucket of car wash soap mix, wash mitt

Talent: ★★

Tab: $5–$10

Tip: Make sure the product selected is compatible with the wheels on your vehicle

Gain: New-looking wheels with little effort

Complementary project: Clean and dress tires

With the availability of so many different types of wheels, a number of easy-to-use wheel cleaners have also become available. As much as wax is a controversy among auto enthusiasts, so are wheel cleaners. Eagle One, Meguiar's, Mothers, and other cosmetic car care manufacturers offer a large assortment of various wheel cleaners. They recognize that almost every type of wheel needs a separate cleaner and no single wheel cleaner is right for every wheel type. Therefore, read labels to distinguish the right wheel cleaner for the type of wheel on your automobile.

Wire mag, anodized aluminum, billet aluminum, polished, and painted wheels need special treatment. Some wheel cleaners claim to be good for all-purpose use. This may be true only because of the weaker nature of the product when compared to some of the specialty wheel cleaners. Basically, an acid mixture is used to rid wheels of stains and dirt. Too strong a mixture will ruin certain types of wheels. Hence, an all-purpose product must be relatively weak. Manufacturers have done their homework and have come up with products that are safe for all wheels. Some contain acid ingredients and others do not.

It is perfectly fine to use an acid-based wheel cleaner one time. Then stay on top of the cleaning problem by washing frequently. Use a toothbrush and soap. Put on a pair of soft cotton gloves and use your fingers as washing tools.

Many specialized chemical wheel cleaners contain hydrofluoric, phosphoric, or oxalic acid. This is the chemical ingredient that dissolves brake dust, oil, and road tar. Before applying any wheel cleaner, be sure the wheels have cooled down after driving. You can help them cool with water spray after they have sat idle for a while. Do not spray cold water on hot wheels!

Instructions advise users to apply cleaner and let it stand for a specific amount of time, generally 30 seconds to two minutes. This is critical. Rinsing too soon won't allow the cleaner enough working time, and leaving it on too long may result in wheel damage. For extra tough grime and brake dust buildup, agitate the cleaner on the wheel with a paintbrush.

Rinsing is just as critical as the amount of time you let the cleaner sit on the wheel. Most instructions will suggest using high-pressure spray as opposed to a thumb over the end of a garden hose. Thorough rinsing is necessary—in fact, you cannot rinse too much. Besides getting acid off the wheel, you should be concerned about removing residue from valve stem threads, lugs nuts, center caps, spoke nipples, and the like. To be on the safe side, plan to wash wheels with regular car wash soap and a wash mitt after chemical cleaner has been thoroughly rinsed off.

In lieu of, or along with, wheel cleaners, seriously consider the use of polishes designed for the material your wheels are made of.

If your wheels are in bad shape and need serious attention, do not rely solely on wheel cleaner products to make them look new. It took time for the wheels to get into their neglected condition, and you can expect it to take some time to get them into like-new condition.

TECHNIQUE 49 WIRE WHEELS

Time: 2 to 4 hours

Tools: Soap and water, wash mitt, paintbrush, toothbrush, chrome polish, wax, cloths or towels

Talent: ★★

Tab: $5–$15

Tip: Have patience; cleaning, polishing, and waxing wire wheels correctly takes time

Gain: New-looking wire wheels

Complementary project: Clean and dress tires; clean and paint fenderwells

Wire Wheel Chemical Cleaner

Some chemical wire wheel cleaners are two-step processes. The first step is a cleaner and the second a neutralizer. Rinse the wheel first with plain water. Next, spray the first-step cleaner liberally on the rim and spokes, avoiding plastic center caps. Allow the mixture to set approximately 30 seconds, and then rinse with water.

Step two, the neutralizer, is applied in the same manner as step one. The neutralizing agents in step two contain water softeners and specific agents that help to prevent streaks, stains, and water spots. Rinse with plenty of water and dry with a clean soft towel.

This type of product, used according to instructions, will make short work of cleaning wire wheels. Light rust, road grime, tar, oil, and brake dust are easily and quickly removed.

Once again, however, one has to wonder about the long-term effects after repeated, frequent applications. Consider using such a cleaner once, if you have to, and then maintain your wheels with a weekly wash using mild soap and water.

Basic Wire Wheel Cleaning

Cleaning spoke wheels is labor intensive. Once a year or so, plan to remove wheels one at a time from your car so that both sides of the wheels can be thoroughly cleaned, polished, and waxed. Ensure the vehicle is safely supported with jack stands while the wheels are removed.

Armed with a pair of soft cotton gloves, wash mitt, paintbrush, and toothbrush, clean each and every spoke and those spaces on the rim between them. The preferred toothbrush has a bend just behind the bristles. It allows better access to tight spots. A paintbrush also works well to clean around nipples and at those points where spokes are positioned close together. With a pair of cotton gloves, you can use your fingers as miniwash mitts, too. Just dip your hands into the bucket of wash soap and massage the spokes clean.

Dish soaps designed to resist water spotting are good. You can also use multipurpose cleaners, like Simple Green. Work up a good lather with the mitt and follow with the paintbrush or your cotton-gloved fingers. Rinse frequently, both to remove soap before it dries and to give you a better look at what you are doing. Don't forget to clean the back of the wheel, if you have it off the car.

Threaded parts of the spoke next to the nipples are common spots for rust. If the toothbrush and soap won't remove rust, try a little chrome polish on the end of a cotton swab. Remove polish with a towel and toothbrush. If rust persists, apply polish with a toothbrush.

Since the backs of these wheels generally receive so little attention, many are found in very dirty shape. Use a wet scouring pad to remove heavy accumulations of road grime, dirt, and brake dust. A scouring pad works well on the spokes, too. Pitted and rusted rims will need polishing with Happich Simichrome or another chrome polish.

Wax the wheel for added protection. This is a good idea for the back of the wheel, too. Apply wax carefully, avoiding buildup at spoke bases. Remove buildup quickly with a cut-off paintbrush. Squirt a little WD-40 into all spoke ends to help reduce future rust problems.

The center cores of knock-off wheels should be lubricated. After a thorough cleaning, be sure to replace lubricant with the proper grease. You can also clean knock-off threads with a toothbrush and a rag. Once a year is not too often to check spokes for tightness. If you know how, tighten them yourself. If not, use the Yellow Pages to find a reputable wheel shop that can true the spokes.

1. After rinsing with plenty of water, spray wire wheels with a good liquid cleaner. Simple Green, Meguiar's Extra, and Gunk all work well. Let the cleaner soak in for a moment before wiping with a wash mitt, sponge, or cloth. This will help to loosen brake dust and dirt.

2. A cloth or small towel may be difficult to maneuver while washing wire wheels. It is difficult to get the cloth into the tight spaces between wires. Opt for a wash mitt or cotton gloves.

3. A small paintbrush works great for reaching inside the wheel past the wires. Be sure to cover the metal band of the paintbrush to eliminate possible scratching.

4. Polish chrome wire wheels as you would any chrome piece. A small sponge or cloth will work best, as either will be more manageable around the wires. Use a paintbrush with the bristles cut to about 1/2 to 3/4 of an inch to help remove polish from around nipples and other tight spots.

TECHNIQUE 50 MAGS AND OTHER SPECIAL WHEELS

Time: 1 to 3 hours

Tools: Soap and water, wash mitt, paintbrush, toothbrush, polish, wax, towels

Talent: ★

Tab: $5–$15

Tip: Place a piece of plastic over the lug nuts before slipping on the lug wrench to prevent scratches

Gain: New-looking wheels

Complementary project: Clean and dress tires; clean and paint fenderwells

Chemical Cleaners

True mag cleaners contain no acid. Instead, they rely on solvents to penetrate the surface and remove accumulations of road grime and brake dust. Since acid is not used, a neutralizer is not needed, making mag cleaners a one-step process.

Apply according to directions. Make sure the wheels are cool to the touch, and rinse them with water. Spray cleaner liberally on the wheel and let it stand for one to two minutes. Rinse thoroughly with a strong water spray.

Be sure to read mag cleaner labels before purchasing. Some cleaners are made for machine-finished and open-pore cast-aluminum wheels, while others are for clear-coated, polished, and painted factory wheels.

If you are not sure what type of wheels are on your car, stop by a wheel shop to find out. Using the wrong cleaner may damage wheels, requiring repolishing or other repairs. Auto parts stores generally carry an ample variety of chemical wheel cleaners.

Use an older wash mitt for extra-dirty wheels. Once a year or so, remove the wheels from the vehicle one by one so you can clean the backs. Ensure the car or truck is safely secured on jack stands.

Wash the wheel first with the mitt. Push it into tight spots along slots and fins. Use a paintbrush liberally. You will be surprised at how well it cleans inside the fin and gridwork patterns. Use a toothbrush around screws and inside slots. Don't be afraid to use your fingers, too. Put on a pair of soft cotton gloves, dip your hands into the wash bucket of soapy water, and massage wheels clean. Your fingers are amazing tools, as they can fit into places the wash mitt can't, and you won't have to worry about scratch hazards.

The backs of some wheels may have to be scrubbed with a plastic-bristled brush to remove heavy concentrations of road grime and brake dust. This harsh cleaning is not good for wheels, as it can cause slight scratches as grit is moved around by the brush. Use this method only as

Cleaning Other Wheels

Painted, polished, clear-coated, cast-aluminum, and steel wheels are cleaned with mild soap and water. After scrubbing the tire, rinse the wash bucket and mix up a fresh solution of car wash soap or liquid dish detergent.

Once a year, or more often if you like, pull the wheels off the vehicle one by one so you can clean the backs. It is amazing how much crud builds up on these surfaces. This is a bigger concern for those wheels that feature wide openings that expose large portions of the backs.

Tires & Wheels   137

2. At first glance, this wheel looks clean. However, look at the corners where the spokes meet the outer rim and you'll notice small pockets of dirt remaining. Spray corners like these with cleaner and agitate it with a paintbrush. Also take note of the dirt and grime buildup on the exposed back of the wheel.

3. A little work with cleaner and a paintbrush got rid of the dirt in the corners. Some light work with a toothbrush removed a lot of the grime buildup on the wheel's exposed back. This wheel should be pulled off so the back can be thoroughly cleaned and protected with a coat of wax. The exposed back should then be cleaned with a toothbrush each time the wheel is washed.

1. After a thorough rinsing, spray liquid cleaner onto wheel surfaces. Let the cleaner soak in for a few moments to give it a chance to loosen up pockets of dirt, brake dust, and road grime. Be sure to get the cleaner into slots and openings for complete coverage. Use a wash mitt to initially wipe down the wheel, and then use a soft scrub brush or paintbrush to agitate the cleaner in and around recessed wheel parts. A paintbrush works well to get into the tight lug nut spaces.

Some inventive detailers like to use their fingers to reach into tight wheel spaces for cleaning. They put on a pair of cotton gloves and use their fingers as mini wash mitts. Gloves protect fingers from scratches and also do a good job of loosening up dirt and brake dust.

needed to remove the really tough stuff. Detailed cleaning should be done afterward, using a soft paintbrush or toothbrush.

Some BMW wheels can be tough to clean. The gridwork pattern includes a lot of pockets that are difficult to reach. Persist with the paintbrush, using plenty of suds from the wash bucket. You may try using a special toothbrush made for people with orthodontic braces. It is a soft brush with bristles shaped like cotton on cotton swabs. Don't force this brush into pockets, as it does have a metal rod supporting the bristles that could scratch the surface.

Chrome-plated steel and polished mag or aluminum wheels are polished with a chrome and mag polish product; such are available from Meguiar's, Mothers, Eagle One, and others at most auto parts stores. Painted wheels are polished with a regular auto paint polish.

All should receive a light coat of wax afterward. Clear-coated wheels must be polished with products specifically designed for them.

For badly neglected chrome wheels, try Happich Simichrome or any of the other brand-name chrome polishes. Extra heavy stains may require three to five applications of polish to completely shine. If need be, use number 0000 steel wool with polish for maximum shining results.

Conscientious detailers worry about scratching nice lug nuts when taking them off for wheel removal. To protect their finish, place a piece of plastic over them before putting on the lug wrench.

TECHNIQUE 51 WHEEL COVERS AND HUBCAPS

Time: 1/2 to 1 hour

Tools: Soap and water, wash mitt, paintbrush, toothbrush, all-purpose cleaner, towels

Talent: ★★

Tab: $5

Tip: Remove wheel covers once a year for thorough cleaning

Gain: New-looking tire and wheel accessories

Complementary project: Clean and dress tires; clean and paint fenderwells

Most wheel covers and hubcaps are strong enough to withstand numerous removals. Even so, you must use care in pulling them. The stock beauty rims and hubcaps on Corvettes bend easily. Do not use tools to pry them loose. Use your fingers, and pull a little at a time, constantly moving around the rim. For those and other types that are fragile, opt to leave them in place for detailed cleaning, rather than take the chance of bending or denting an edge.

Painted wheels can suffer chipped paint during the removal of wheel covers. Use a wide-bladed tool to pry loose an edge. Then place a thin piece of cardboard or a towel between the tool and the rim edge. This will help reduce metal-to-metal wear and prevent paint chips.

Just as much care should be given to putting covers back on wheels as to taking them off. Be sure the valve stem lines up with the hole in the cover. Misaligned, the sharp edge of the cover may cut the valve stem to cause a slow leak. Use your hands to get the cover in place and secure it with a rubber mallet. Since many wheel covers host plenty of tangs with sharp edges, consider wearing a pair of gloves while working with them.

Wheel covers and hubcaps off the vehicle are cleaned just like any other part. Use soap and water and a mild liquid cleaner to remove dirt and grime accumulations. A paintbrush and toothbrush work well for getting into the tight spots. As needed, spray wheel covers and hubcaps with an all-purpose cleaner. Agitate the cleaner with a paintbrush to loosen grime buildup. A toothbrush works well on stubborn spots in grooves and recesses.

Chrome wheel covers and hubcaps are polished just like other chrome parts. Those made of different metallic materials should be polished with a product safe for all metals. Apply polish with a soft cloth and buff clean with another. Apply a light coat of wax the same way.

Those covers and hubcaps in need of paint are fixed up with the use of a fine artist's paintbrush. Check with your local autobody paint and supply store for the correct paint colors. Mask as needed with quality masking tape; cheap masking tape from variety stores tends to let paint seep through edges and is generally not well suited for automotive paint projects. Make sure you have the proper solvent available for brush cleaning and general cleanup.

Clean wheel covers and hubcaps as you would wheels. Start out with car wash soap and water, spray them with an all-purpose cleaner, and agitate with a soft brush. Rinse and dry.

TECHNIQUE 52 PAINTING WHEELS

Time: 1 to 2 hours

Tools: Paint, paint block, masking tape and paper, thinner for cleaning up

Talent: ★

Tab: $5–$25

Tip: Be certain wheels are clean and dry before painting

Gain: New-looking painted wheels

Complementary project: Clean and dress tires; clean and paint fenderwells

Stock-painted wheels can be repainted. The best way is to remove the tire and strip old paint to bare metal. Sand and prime as necessary and then paint. An autobody paint and supply store will mix the proper paint, matching the original color, or help you to create a custom hue. Enamel paint is the most common because of its durability; unlike lacquer, enamel does not have to be buffed out once it dries. Have paint and an artist's fine paintbrush on hand when remounting the tire. This will enable you to touch up any chips that occur during the process.

Wheels may also be painted with the tire attached, either while on or off the vehicle. The end result may not be as nice as a wheel that has been stripped, sanded, and painted, but with care a wheel's appearance can be greatly improved.

If using a spray paint, prevent overspray on the tire with masking tape, grease, or a paint block. Masking tape will not stick to a tire that has been recently freshened with dressing. You can carefully mask next to the rim with quarter-inch masking tape and cover that with a wider strip three-quarters to one inch wide. Cover the rest of the tire with masking paper or double-thick newspaper, taping it to the strip of tape already on the tire. The first masking strip you place next to the rim is the most critical. Take your time and try to place tape behind the rim edge as much as possible.

A film of grease can be placed along the tire next to the rim. Paint will not stick to grease. This is a messy chore and you must not get any grease on the rim, as paint will not stick to that grease, either.

A paint block can be made out of a piece of thin cardboard. The bottom of a shoebox is perfect. As you paint around the wheel, keep an edge of the paint block tucked closely between the tire and rim edge. Be sure the paint block is in position before you start painting. The block will have to be held in one hand while you paint with the other. Overspray should be minimized. If any paint gets on the tire, carefully remove it with lacquer thinner.

Chips on painted wheels are repaired much like chips on body paint. Dab paint on the chip. Let it dry and do it again. Allow the paint to build up slightly higher than the surrounding surface. After a week of drying, mask off the chip and sand it smooth with number 600 wet-and-dry sandpaper. Remove the masking tape and then polish to a fine finish. Wait a month before applying wax.

Caring for Lug Nuts

Before replacing wheels, take a minute to service the lugs. Remove with a rag and WD-40 on a toothbrush.

Chrome and painted lug nuts may get scratched while being tightened with a lug wrench. To prevent scratches, place a piece of plastic over the lug nut before placing the wrench over it. One side of a heavy-duty freezer bag, such as a Ziploc Bag, works well.

Masking a tire in preparation for wheel painting is somewhat tedious, because the wheel is round. Plan on using a lot of tape. Clean and dry the wheel thoroughly and sand if the surface is rough or the paint is flaking or chipped. Apply two or three light coats of paint.

Chapter 9
Finishing Touches

Take care of all those sometimes-forgotten projects that, done properly, will elevate your vehicle to perfection.

Detailing Techniques

This section describes how to take care of a lot of the little things that may be overlooked while trying to make an automobile look, feel, and smell as good as new, or even better.

Although these topics may not fit well into the other categories, they cover points you should consider in your detailing program. Fit them in as you deem appropriate.

This classic vintage automobile has been thoroughly and meticulously detailed and doesn't necessarily look overly brilliant and stunning. It does, however, appear fresh, clean, and crisp to stand tall and be admired.

A detailed interior with everything neat and clean. Seats have been treated to a conditioner with the excess buffed off to leave behind an appearance of luxury.

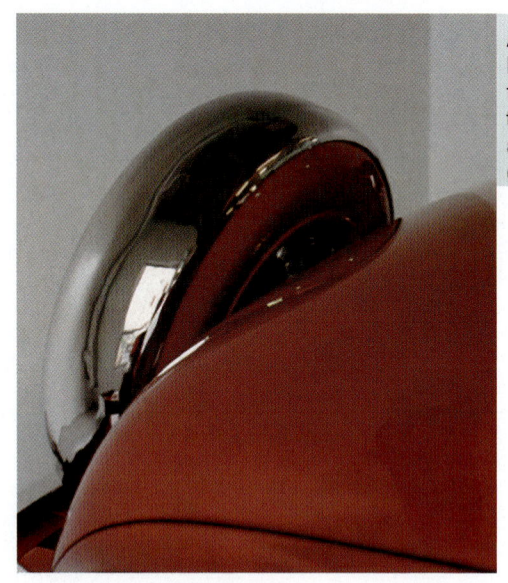

All of the little details have been addressed. No finger smudges on paint, the chrome is perfect, and no hint of dry polish or wax anywhere.

From top to bottom and front to back, this vintage classic challenged the detailer with all kinds of nooks, crannies, and accessories that had to be cleaned and shined. You can see that any part left unfinished would negatively impact the overall beauty of this fine automobile.

TECHNIQUE 53 TRUNK REJUVENATION

Time: 1 hour; 4–8 for total restoration

Tools: Vacuum, cleaner, towels, polish and wax. For restorations, wire brush, sandpaper, primer, spatter paint, masking tape, and paper

Talent: ★–★★★★

Tab: $5–$75

Tip: Total trunk restoration may require a little work over the course of a couple of weekends

Gain: Trunk space that looks brand new

Complementary project: Service the jacking equipment

1. This neglected trunk features light, medium, and heavy rust accumulations. A portion of the floor has rusted through. In lieu of replacing the trunk floor with new metal, the detailer will treat the rust, repair the damaged metal with fiberglass, and then coat the entire space with GM Spatter Paint. The pole seen on the right side of the trunk is holding up the trunk lid, as new trunk lid support rods have yet to be installed.

2. Loose flakes of rust and other chunks of dirt and debris are loosened up with a wire brush. A small wire brush works well to get into the corners and tight spaces.

3. All of the loosened debris must be removed. A dry paintbrush helps to gather the big stuff, and a vacuum cleaner will remove the remaining dust.

4. The detailer for this project will use these products to prepare the metal for spatter paint. Marine Clean is used to wash away dirt, oil, and grime. Metal Ready will remove the remaining rust and POR-15 will coat the surface to prevent rust from coming back. When using products like these, be sure you work in a well-ventilated area, and consider wearing a mask along with safety goggles and rubber gloves.

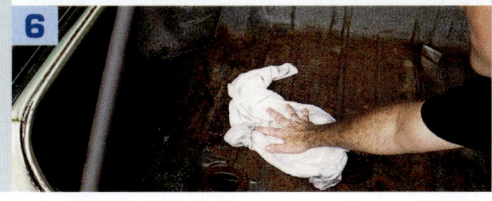

Trunk Cleaning

If the trunk you are about to clean is dirtier than anything you have ever seen, seriously consider pulling the rubber drain plugs and using plenty of soap and water and the plastic-bristled brush to scrub. You will have to dry it afterward. A wet/dry vacuum cleaner helps, along with a few towels. Use caution with the garden hose. Don't get the fiberboard back part of the rear seat wet, either.

Sometimes you'll find the floorboard of the trunk in rusty condition. In extreme cases, you may have to use a putty knife and wire brush to scrape off the big stuff. Then use a fine-grit sandpaper to remove remaining rust. A damp towel and vacuum cleaner will pick up residue.

For trunks in good shape, use a towel sprayed with Simple Green. Follow up with a folded towel, damp with water on one side and dry on the other. Give the trunk a close inspection and clean all of it, including the locking mechanism, brackets for the taillights, and bottom side of the trunk lid, as well as the perimeter edges of the trunk lid and trunk opening.

5. Mixed with water according to label instructions, Marine Clean is sprayed liberally on the surface and scrubbed with a brush. Drain plugs should be pulled so the trunk can be rinsed with water. Be sure the runoff from this step will not cause problems on the driveway or other surface the vehicle is parked on.

6. After rinsing, use a towel to dry the trunk space. Go over spots another time if dirt or oil still lingers.

Detailing Techniques

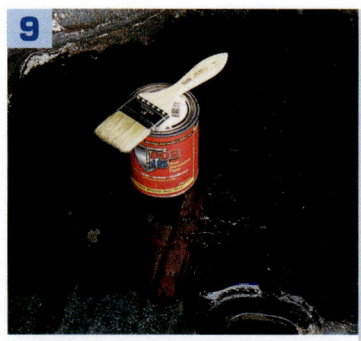

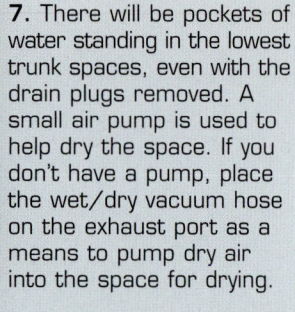

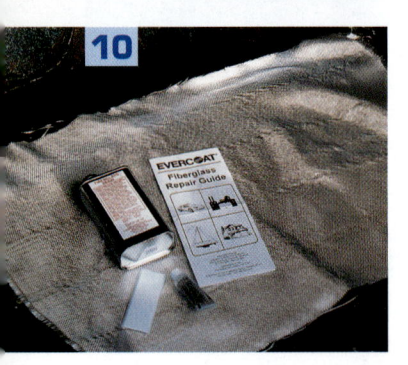

7. There will be pockets of water standing in the lowest trunk spaces, even with the drain plugs removed. A small air pump is used to help dry the space. If you don't have a pump, place the wet/dry vacuum hose on the exhaust port as a means to pump dry air into the space for drying.

8. After the space has been thoroughly cleaned and dried, Metal Ready is sprayed on the surface according to label instructions. For this extra heavy rust treatment job, the detailer sprayed on a new coat of Metal Ready every half-hour for about 1-1/2 hours. After that length of time, no more rust-removing progress was noticed and the Metal Ready was allowed to dry.

9. This area of metal damage was coated with POR-15 first in preparation for fiberglass. It was allowed to dry overnight. After the fiberglass repair is made, the entire trunk space will be coated with POR-15 in preparation for spatter paint.

10. A fiberglass cloth sheet will be cut to fit over the repair area and then coated with resin. These materials are generally available at auto parts stores in kit form. The resin requires a catalyst in order to make it harden. Be sure to follow mixing instructions carefully. If mixed improperly, resin could harden much too fast and adversely effect the proper repair.

11. The fiberglass sheet was cut with scissors to fit into the place needed. Notice that the fiberglass has been positioned away from the drain hole so that the cover will fit properly when put back in place.

12. Fiberglass has been coated with resin and will be allowed to dry. Resin was placed on top of the fiberglass with a paintbrush and smoothed with a small plastic squeegee.

Trunk Rejuvenation

Nicely painted trunks should be lightly polished and waxed. You just can't go wrong cleaning, polishing, and waxing all painted surfaces.

Vintage American cars are generally complemented with large trunks. Many have been painted with spatter paint, a slightly rough coating that consists of gray paint dotted with white specks. For trunks like these, especially the ones that require rust removal, clean and sand as needed. Then apply two coats of a rust-inhibiting paint or primer. Allow it to dry and then spray on a coat of Zolotone or other trunk spatter paint. These paints come in different colors and are available at local auto parts stores and autobody paint and supply houses. Mask as needed, using good masking tape, masking paper, or double-thick newspaper and towels, as you deem necessary. The end result should provide your car with a trunk space that looks unused.

If you don't like the original look of spatter paint, use whatever type and color of paint desired. Be sure the surface is thoroughly sanded smooth and free of sanding dust. If not, flaws will be most noticeable. Follow the directions on the paint label and wait the appropriate time to let paint dry before reassembling the trunk.

The Process

Once the trunk space has been emptied, survey the floor and sides to determine how much work is needed to make it look new. Severe rust problems, with holes all the way through the metal, may require new metal installations. For those that suffer mild to medium rust, you should treat them first before starting any painting endeavor.

After removing the drain plugs, scrape off the loose rust with a wire brush, using a putty knife as necessary. Detailers have enjoyed good results using Marine Clean to scrub off old dirt, oil, grime, and the like. Rinse with water and use a wet/dry vacuum to get the water out of the pockets in the lower sections next to the quarter panels. Pat the rest of it dry with a towel.

Once the trunk space is clean and dry, apply a coat of rust neutralizer, like The Eastwood Company's Rust Encapsulator or POR-15's Metal Ready. Follow instructions carefully to ensure a good job. More than one application may be necessary, and you must wait until each coat is dry before applying another. Depending on the neutralizer used, you may need to follow with an application of rust inhibitor before painting. Eastwood's Rust Encapsulator does not need to be top coated, while others require a rust inhibitor, like POR-15. Be sure to follow label directions exactly, work in a well-ventilated area, and wear gloves.

Finishing Touches

13. A wire bundle and openings along the panel behind the back seat were masked off to protect them during the application of POR-15. The first coat was allowed to dry for about five hours before the application of a second. That was then allowed to dry overnight. Masking will remain in place until the entire space has been coated with spatter paint.

14. Once POR-15 is completely dry, the surface is very lightly sanded with medium-grit Scotch Synthetic Steel Wool. Sanding is done lightly—not enough to break down the surface coat, but just enough to allow the best adhesion of the trunk spatter paint.

15. Although trunk spatter paint is available at auto parts stores, you may want to check with a dealership to ensure the product you use is factory original. This GM product is water-based and must be covered with a topcoat after it has dried according to directions.

16. Body side emblems were removed and their stud openings were covered with masking tape on the outside to ensure adequate spatter paint coverage. Towels are placed over the areas around the trunk to protect them from overspray.

17. Be sure to remember to clean, prepare, and paint the covers that were removed from the trunk initially. Consider buying new rubber drain plugs, too.

18. Some work over the course of two weekends resulted in this new-looking trunk. A project like this will require plenty of time, because materials must be allowed to dry before additional coatings are applied. Notice that the fiberglass repair is virtually invisible.

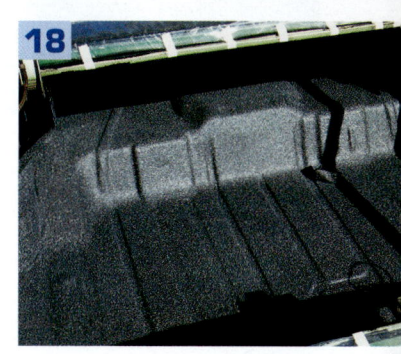

Once the trunk space has been cleaned and rust treatment completed, you are ready for paint. Trunk spatter paint is available at most auto dealerships, auto parts stores, autobody paint and supply outlets, and through The Eastwood Company. Several colors are available; you can research service manuals to determine the stock color of your classic car. A dealership may be helpful; so may autobody paint and supply outlets and car clubs.

Remove side marker lights and other obstacles in the trunk in preparation for spatter paint. Use quality masking tape and paper to cover everything you do not want painted. Take your time, and ensure a good masking job. If you elect to use newspaper in lieu of masking paper, make sure you put it on double thick; one sheet of newspaper will allow paint to bleed through.

Have at least two cans of spatter paint available for a full trunk paint job. Three may be better for those that were cleaned down to bare metal. Apply according to the label directions. Then, after the spatter paint has dried, about 24 hours, apply a clear coat over the top. This is especially important when using water-based spatter paints.

When the paint has dried and you are satisfied with the job, start putting things back together. If the drain plugs were damaged during their removal, remember to get new ones. They should be available through a local dealership.

If you have forgotten how the spare tire and jacking equipment are supposed to fit in their prescribed spaces, check your notes, the pictures you took at the beginning of the project, or the instruction sticker on the trunk lid. If the sticker is missing, you might have to contact a dealership, someone in a car club with auto makes and models like yours, or one of the classic auto parts suppliers to see if OEM replacement stickers are available. Almost all original manufacturer stickers and labels are easy to find.

Many car people wrap jacking equipment in towels before stowing in the trunk. Towels protect the tools and help to lessen road noise by preventing vibration. Some serious folks carry a cleaning kit, a tool kit with some minor parts, a car cover, and an assortment of other automotive stuff in their trunks. It comes in very handy on long and extended road trips. A pair of cotton gloves and a small roll of paper towels come in handy, too. Gloves may be used for changing tires and paper towels for removing bird and insect droppings found on your car while away from home. Simple Green even offers a plastic bottle filled with moist towelettes for cleaning hands and other things.

TECHNIQUE 54 PARTS REPLACEMENT

Time: 1/2 to 1-plus hours

Tools: Replacement floor mats, pedal covers, decals, emblems, and the like

Talent: ★

Tab: $5–$75

Tip: Make a list of all the parts you need and get them as your budget allows

Gain: An exceptionally fine auto detail

Complementary project: Wash and dry the entire vehicle

How do the floor mats look? Do they match? Do they fit your car correctly? Are they the right color scheme? Why lessen the impact of a professionally detailed interior with a set of cheap, ugly floor mats? Go out and purchase a set that fits the vehicle and blends with the overall appearance of the interior. They cost from $3 to $35.

The same holds true for worn pedal rubbers and steering wheel covers. These parts are inexpensive, they last a long time, and they provide the interior with a feeling of completion.

Under the hood, look at the battery and the plastic containers for windshield washer fluid and radiator overfill. A battery that looks old and worn probably is. Why clean plastic bottles if they are cracked and serve no useful purpose? Get a good battery that will serve your automobile well for years. If you can't find a new plastic container, go to a wrecking yard and buy a used one.

If both heater hoses are black, and one bursts, don't replace it with a red one. Actually, if one broke, the other is probably about ready to, so replace them both. Ditto for radiator and vacuum hoses.

The same approach holds for ignition wires. If one goes bad, replace the entire set. If you really want to make a car person laugh, replace a faulty black spark plug wire with a yellow one. Ignition wire sets range in price from $25 to $50 and up, depending on the style and manufacturer. Most engines are equipped with guides for plug wires. They are usually brackets attached to valve cover bolts. The brackets support plastic pieces that hold each wire in place, keeping the wires from dangling over the edge of the valve covers. Make use of such brackets to keep wires neat and aligned.

Factory engine stickers add originality. Replace worn and torn stickers with new ones. Check with auto parts stores, dealerships, autobody paint and supply houses, and the Internet for availability. Radiator caps, fan belts, valve cover gaskets, and battery cables all fit into this category. These parts are inexpensive and seldom have to be replaced. Spend an extra $50 or $60 (more if you need a battery) and make the engine compartment really stand tall.

Look at the wheel hubs on the exterior. Are they adorned with a factory sticker in the center? Are the stickers in good shape? If not, talk to a local dealership to see if they are available. You can also

Don't let unsightly small parts ruin an otherwise great detail job. Replace things like this worn-out door handle insert piece. It is too far gone to clean up and rejuvenate.

Finishing Touches 147

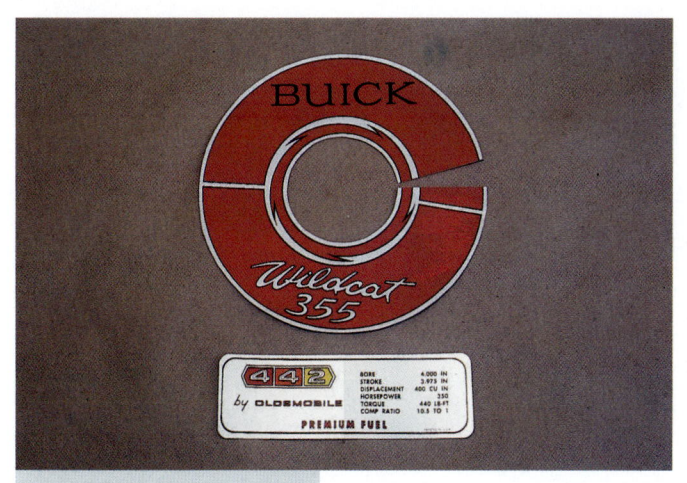

Above Left: You may be surprised to learn how many small parts are available for automobiles of virtually every year, make, and model. Stickers are still made for a lot of different parts on automobiles. Wheel hubs, door edges, sun visors, fuel filler areas, and other spots are commonly adorned with stickers from the factory.

Above Right: Battery trays are rather inexpensive and easy to replace. Be sure to remove the battery while detailing the engine compartment to ensure the battery tray is in good shape and not rusted through.

check the Internet for parts houses that specialize in older vehicle stickers and emblems. All stock stickers, emblems, and decals belong on the vehicle. You should make a sincere effort to find and install them when the old ones are looking bad or have pieces missing.

Original Equipment Decal Replacement

Stock stickers, decals, and emblems for all parts of vehicles are available through a number of different outlets. Some auto parts stores and dealerships can order them for you, and you can surf the Internet to find specialty parts houses that will send you catalogs or allow you to purchase online.

These pieces are especially important for those vehicles that are being detailed as part of a restoration project in preparation for car shows and other exhibition events. Not nearly as important for custom cars, hot rods, and other specialty rigs, original equipment stickers, emblems, and decals are a must for those vehicles being brought back to their original condition.

You should be able to locate stickers, emblems, and decals for air cleaner housings, hoods, radiator supports, doorjambs, trunk lids, wheel covers, and just about anything else that came with the make and model vehicle you are restoring. Many times you can purchase a complete package to cover everything that came with the vehicle from the factory. This is a huge concern for anyone contemplating a future Concours d'Elegance competition.

Old stickers, emblems, and decals must be carefully removed. Use an adhesive remover for complete cleaning. Then take your time installing the new pieces. Ensure that they are going on in the correct spot on the vehicle.

Make sure the spot where you are planning to install the piece is clean and dry. Then peel back just a corner of the backing to start out. This small section is attached first to give you some wiggle room in maneuvering it into place perfectly. You can put it on and then take it off once or twice to ensure the sticker, emblem, or decal is going on straight. Once you have it started and it is true, go ahead and slowly peel back the rest of the backing and slowly attach the rest of the piece. Use your finger and a soft cloth to press the piece down as you go. This will help to smooth out the material and also get rid of any air bubbles.

Be aware that not all stickers, emblems, and decals will allow for wiggle room! Some are pretty unforgiving and must be put on straight the first time. The adhesive and material from which these pieces are made will not allow you to touch one part to the surface and pull it back off without suffering damage. So be careful. Take your time to plan out the task, and do it right on the first try.

TECHNIQUE 55 JUST NICE OR NEW CAR SMELL

Time: 1/4 hour

Tools: Air freshener or deodorizer

Talent: ★

Tab: $2–$5

Tip: Remove the cause of bad odors before using an air freshener or deodorizer

Gain: Pleasantly scented auto interior

Complementary project: Clean the interior

Air fresheners and deodorizers are sold in solid and liquid form. The solid types come shaped like trees, rainbows, and such. Hang them from a knob or a lever under the dash or hide them under a seat. The deodorizing power is supposed to be unleashed a small amount at a time, determined by the degree to which the package is opened.

For severe odor problems, first discover and remove the cause. Be sure to completely clean the area around the odor-causing thing, too. Then use a deodorant full force. Open the package, toss the element under the front seat, and lock the car for three to four days. This will work, provided you removed the cause of the odor first.

Liquid air fresheners and deodorizers may be sprayed directly on carpet. It is recommended you do this under a seat, however, just in case there is something in the product that could stain or fade carpet. Detailers have also had good luck spraying air fresheners inside air vents and ducts.

"New car smell" fragrances are available, as are all sorts of others. You will have to try them yourself to see if they meet your expectations.

The best way to make your car smell good is to keep it clean. There is nothing better than the smell of clean. It is impossible to describe but easy to detect. Keep the interior vacuumed, wipe it down frequently, and remove the cause of bad odors immediately. Remember, an air freshener or deodorant may mask an unpleasant odor, but it will not get rid of it if the cause lingers. You can enhance and extend the clean smell with lemon-scented Pledge or by adding a bit of Lysol to cleaning solutions and carpet shampoo.

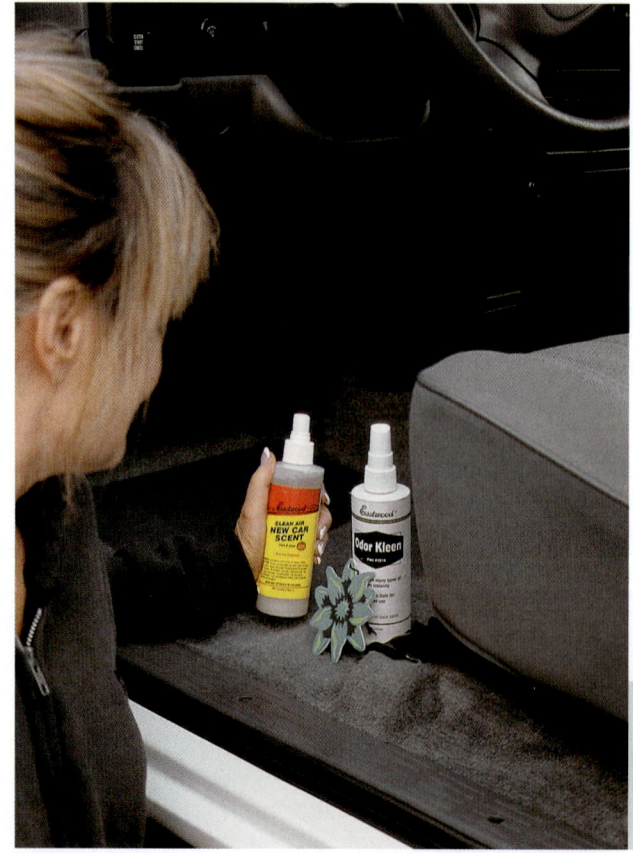

Air fresheners are available in both liquid sprays and solids. Under front seats is a favorite spot for application of the sprays and placement of the solids.

Finishing Touches 149

TECHNIQUE 56 BRA CARE AND CAR COVERS

Time: 1/2 to 1 hour

Tools: Soap and water, wash mitt, vinyl/rubber protectant, soft dry towel

Talent: ★

Tab: $5

Tip: Regularly remove bras to clean the car body underneath

Gain: Protection of the paint finish

Complementary project: Polish and wax the area under the bra

Protecting a beautiful detailed automobile is important. Car enthusiasts have mixed emotions about bras. They protect paint against chips from road hazards. At the same time, they trap rainwater and minute pieces of grit. As a bra vibrates in the wind, grit is rubbed against underlying paint to cause scratches. If you deem it necessary, put on a bra while traveling a particularly rough road. Remove it when you are past the hazards.

Bras themselves are maintained in much the same way as soft interior vinyl. Break loose and whisk away lint and grit from the inner surface with a soft brush. Clean the outside with a wash mitt and car wash soap, using a soft brush as necessary to remove buildup and bug residue. Hang the bra on a clothesline or other suitable spot and allow it to dry.

Afterward, apply a light coat of vinyl dressing and be sure to buff off the excess. Make sure it is dry before storing or replacing on the vehicle.

It is important that bras be attached to host vehicles tightly. Loose straps and other parts that are not fastened securely will flap in the breeze and eventually cause paint blemishes. Take a look under the bra regularly to see if grit and other debris are accumulating. If so, it is time to remove the bra and wash it and the car. Be certain there is a good coat of wax on the car under the bra, too.

It is important to clean the area under a bra regularly. Fine particles of dirt and grit can get lodged behind it to cause minor scratch blemishes. Clean the bra with soap and water, and hang it up to air dry. Be sure the surface protected by a bra has a good coat of wax.

Car Covers

Car covers do an excellent job of protecting paint and interiors. Most important in the summer, good car covers protect paint from the scorching rays of the sun. They also keep the sun's ultraviolet rays from penetrating interiors to dry vinyl and leather upholstery and fade fabrics.

Plastic car covers are worse than no cover at all. Heat is trapped inside with no place to go. Your car needs a cover that breathes. Cot-

Detailing Techniques

Purchase a quality car cover that breathes. One of the worst things you can do is cover an automobile with plastic that will trap moisture and eventually cause mildew problems.

ton car covers are favorites among serious car people. Beverly Hills Motoring Accessories carries an excellent assortment of them.

There are no real tricks to putting on a car cover or taking one off. The key is to never let it touch the ground, where it can pick up grit or grime. You might need some help putting a car cover on the first time. After that, taking it off and putting it back on is a one-person operation.

Imagine that the cover is on your car. Loosen one side and fold it to the middle. Do the same with the other side; now the folded cover is lying lengthwise along the hood, top, and trunk. Next, fold the cover into thirds. Fold the far front up to the middle of the hood. Then, starting at the rear, fold the cover all the way to the hood.

When you cover the car, simply place the folded cover on the hood. Unfold to the rear, then to the front, followed by each side. Tuck it around bumpers and mirrors.

Avoid sliding a cover on the car. The slightest grit will cause scratching. If you think it is going to rain, keep the cover off. Rain won't hurt the car; a wet, moldy car cover will.

If you must park your special car in the sun on a regular basis and don't have a cover, try alternating the car's position. Park in different spots, allowing the driver's side to catch most of the sunlight one day, the passenger side the next. If only one parking space is available, back the car in one day and pull in straight the next.

Finishing Touches

TECHNIQUE 57 CONCOURS D'ELEGANCE CONSIDERATIONS

Time: As much as you want

Tools: Clean soft towels, cleaner, Lexol, cotton swabs, paintbrush, toothbrush, tire dressing

Talent: ★★-★★★★★

Tab: $25-plus

Tip: Concours d'Elegance is the ultimate in automotive detailing

Gain: Perfect auto detailing

Complementary project: None

Concours d'Elegance is an event in which owners of restored and highly maintained automobiles compete to see which entry is the cleanest and most original. The amount of time these competitors spend detailing their fine automobiles is staggering. Their efforts are truly labors of love for both the automobile and the competition.

It has been said that a brand-new car, fresh from the showroom floor, might score 150 points out of a possible total of 300. With most winning entries in serious competitions scoring in the 295 to 297 range, it might be safe to say that brand-new cars are only half as perfect as concours winners!

Concours events are set up differently for all makes and models and for various reasons. The ultimate and most serious events are as thorough as you could possibly imagine. Some competitors have spent entire days just on wheels and tires. Others have spent two hours bringing up the perfect degree of gloss on a cast-plastic air cleaner element. To say that concours enthusiasts and winners are perfectionists is truly an understatement.

From this angle, a concours judge would look for spotted glass, dirt along the window trim, smudges on the trunk lid paint, hints of dry polish or wax around the taillight lenses, dirty fenderwells and rear underbody, dirt caked up around the trunk key-lock, dirty trunk edges, and grubby-looking tires and wheels.

Detailing for a Concours Event

An automobile generically detailed at a quality shop will generally score in the middle of the pack at a concours event. The exceptions are those prepared by select detailers with concours experience. Their fees will range from hundreds to thousands of dollars, depending on the condition of the vehicle and the type of competition that will be entered.

For your automobile to be a winner, you must research everything you can get your hands on that pertains to your car's make and model. Study shop manuals and parts books with emphasis on blowups of parts and assemblies to see how they are put together, how the cotter pins are bent, and so on. Join a car club and become part of the support group for members active in concours events. Most of all, be prepared to spend lots of time cleaning, polishing, researching, and detailing.

Detailing for a concours event is different from detailing for the street. For concours, the car must be perfect for an 8- to 10-hour period. Long-lasting effects are not as important. For that reason, some of the detailing tips that follow may seem contrary to what has already been presented.

Tips for the Interior

For the first-time detail on a potential concours car, the interior should be stripped. The degree to which it is dismantled depends on the age of the car. Older models may require removal of door panels and trim to allow for cleaning and dust removal.

After a thorough vacuuming, you may need to use scissors to trim carpet nap that doesn't conform to the proper level. You'll need to inspect upholstery for loose stitching or threads and trim accordingly.

Shampooing the carpet may be acceptable for a first-time detail, but concours people worry about shrinkage and ever so slight color fading. Use the least amount of water possible and consider a very light application of the properly colored carpet dye to bring back the original tint.

Concours detailers prefer to bring as little as possible into the interior during cleaning, including water and liquids. Some have found that Tuff Stuff is a good overall cleaner. They like the foam because it contains less liquid than other cleaners and it doesn't tend to fade colors. Many use dry cotton swabs to remove dust in grooves along the dash, although some do spray the swab (outside the car) with Pledge for added dust absorption.

Never use an all-purpose polypenetrant dressing on the interior. The silicones eventually adhere to almost everything and cause fisheyes on anything you try to repaint. Some detailers prefer to use Lexol exclusively. Three days before judging, they will apply a light coat to leather and vinyl. This time span allows the conditioner to soak in and fade to a rich appearance without extra-high gloss. The three-day period also allows vapors to dissipate, which helps glass to stay clean for a longer time.

All interior metal parts on a concours winner must be free of paint chips. Brake and clutch rods, seat brackets, and even the nuts and bolts securing the seat should be sanded and painted as needed. Metal dashboards should be polished with a mild glaze, such as Meguiar's Number 7. During cleaning and polishing, check to see that the heads of trim screws are in line to point in the same direction. Use a toothpick to remove dirt or polish inside the slots.

Some concours winners have used water and newspaper to clean glass. Others prefer regular glass cleaners. Take your time and clean small sections, never letting any of the cleaner dry on the glass. Before an event, clean each window three or four times and use a clean cloth each time. At that, for concours judging, the insides of windows will only stay clean for 30 to 45 minutes with the car doors closed. This is due to the vapors emitted by interior vinyl parts.

Tips for the Exterior

Serious concours competitors have been known to spend up to 18 hours polishing paint with Meguiar's Number 7. And this is on vehicles that already look terrific! Totally by hand, every swirl and hint of spider webbing is polished out. To fill and remove scratches (scuffs), they apply glaze in any direction necessary. For final applications, polish is applied and buffed off in a straight back-and-forth motion.

Many concours people do not use carnauba wax protection on show cars. They feel that wax builds up and yellows over time. Instead, they polish with a glaze and keep the car out of the sun and the elements, usually in a separate heated garage. Minute swirls and spider webbing remain invisible for about 10 days. Preparing for the next event requires another treatment of the same intensity.

The degree of exterior paint gloss is another important factor noticed by concours judges. The high-gloss detailed look does not fare well. Orange peel (a rough paint surface that looks like the texture of orange rinds) is also a factor considered by judges. If a particular model came from the factory with a certain degree of orange peel, points are deducted if the car owner polished paint to perfection, flattening the original orange peel.

Chrome is polished as it is with any automobile. Some prefer to protect it with a coat of wax. For shows, some competitors apply a light coat of Windex. It brings up a brilliant shine without an oily film.

Meguiar's Number 40 has provided good results on exterior rubber. It is applied to every piece of rubber on the exterior, including tires, bumper guards, and wiper blades. A cotton swab is used to carefully wipe it on rubber next to paint.

Since long-lasting qualities are not the main concern for concours cars, Lexol has been used successfully as a dressing for vinyl tops. It makes the top look fresh without the glossy look silicone dressings exhibit, and it will last through a four- to five-hour show.

Finishing Touches

Concours events for different makes and models generally have different sets of rules. Some require undercarriage inspections and judging, while others promote vehicle operation through slaloms and other courses. All concours events, though, award honors to automobiles in pristine condition, just like this one.

Around the entire exterior of a concours winner, you will notice strict attention to detail. Screw heads will point in the same direction, swirl marks will be nonexistent, and you won't find a stray piece of dried polish anywhere. Glass will be perfectly clear, and tires will look better than new.

Tips for the Engine Compartment

In the engine compartment, judges have checked the back of distributor cap clips for scratches and have probed deep into valleys on manifolds with cotton swabs looking for dirt and grease. Judges have also checked spark plug wires for correct curvature coming out of spark control boxes and proper light reflection bouncing off painted inner fenderwells.

Some competitors like the results they get with Gunk. The product cuts grease and does not induce further damage. As with all cleaners and solvents, it is used sparingly.

All clamps, screws, and bolts that can be aligned should be. The factory original marks should also be maintained. These are the light paint marks made by factory inspectors to denote that the specific part was installed correctly and torqued to specifications.

Paying close attention to detail, you must remove any hint of rust from the edges of clips and brackets and repaint the correct factory color. Serious competitors also look at the radiator. They spend hours ensuring that debris is removed and all of the fins are aligned correctly.

Virtually every square inch of the engine compartment is inspected and then cleaned, polished, and made perfect. Even the hoses and wires are buffed to just the right degree of gloss using Meguiar's Number 40.

Tips for the Underbody

Only for the most serious of concours competitions, the bottom of a vehicle, front and rear combined, can be worth 70 points. Winners spend many hours on their backs, cleaning and polishing undercarriages until you could eat off them. Gunk has been found to remove Cosmolene easily and with no ill results. Cosmolene is a waxlike protectant sprayed on metal parts to protect against rust when vehicles must be transported by ship across the ocean. For other hard-to-remove substances, try solvents and, at times, carefully planned applications of lacquer thinner.

Pledge furniture polish works very well to shine the fenderwells and underbody. Use the unscented type—the lemon-scented Pledge attracts bees, which disturb the judges and leave droppings on the paint.

A winning concours underbody will be painted and undercoated the way the factory intended it to be. Determining the correct color and texture of the undercoat may require extensive research in shop manuals and other sources. To locate information sources, check with dealers, car clubs, friends, other concours enthusiasts, and the Internet.

Final Thoughts about Concours

Please don't let this discussion about concours discourage you from entering the concours arena. You can explore a number of different concours events, each with its own rules and guidelines. Some of the descriptions brought up here are reflections of the fiercest types of competition. Plenty of other concours shows happen every year without the need to clean undercarriages or focus so intently on engine compartments. Many events are held for charity purposes, and entrants are encouraged to drive their cars through slaloms and other courses. Check around your local area to see what types of shows are put on during the spring and summer months, and learn how you can become a part of the fun.

TECHNIQUE 58 FINAL INSPECTION

Time: 1/2 to 1 hour

Tools: Clean soft towels, cut-off paintbrush, dressing cloth, window cleaner

Talent: ★★

Tab: $5–$15

Tip: Pull the vehicle out into the sunlight

Gain: Enhance your overall detailing efforts

Complementary project: Dust off the vehicle

Under artificial lighting, a detailed automobile may look perfect. In sunlight, the perception is much different. The sun will readily reveal smeared windows, swirls, lint, polish and wax residue, and a lot more. Park the vehicle in a dry sunny spot. Have your cleaning gear at the ready, including the dressing-soaked cloth, cut-off paintbrush, and clean towels.

Interior

Open all of the doors and sit in the driver's seat; place a towel on the floor if your shoes are not completely clean. Check the dash for spots missed by dressing. There should be no need to spray additional dressing, as wiping with the dressing cloth will bring up some excess from the surrounding vinyl.

Are the gauges clean? Check for smeared dressing marks, lint, and dust. Vents can be touched up with cotton swabs. Spray swabs with a dab of household furniture polish, like Pledge. This makes the swab more dust-absorbent. Afterward, adjust vents so they point in the same direction.

Check the headliner and rearview mirror, front and back. Look in the corners of the windshield for dirt and around the mirror base for smears. The steering wheel should be squeaky clean, including grooves near the horn button and along the steering column.

Next, check carpet for lint and stains. A dry spot remover will work fine. Worn patches on carpet can be covered with rubber inserts at the upholstery shop. Secure loose carpet using double-backed tape. Trim loose upholstery threads and carpet fibers with a pair of scissors.

Put tension on the seats to spread pleats, looking for lint and grit. Do the same thing to the bead and the area between the cushion and the backrest. Check the back seat area with the same intensity.

Finally, look at the door panels. All scuffs and marks should be gone. If not, clean and dress again. Check door handles and pouches, if your car is so equipped. Doorjambs must be free of dirt and grease. Squirt a little WD-40 on hinges and door latches for lubrication. Wipe off the excess. Are the door moldings in good shape?

Exterior

Open the hood and trunk lid. Look for and remove polish and wax residue on the edges of these parts, including the area next to the windshield and rear window. The lip drain and molding for the trunk lid must also be clean and dust-free.

Inspect decals, emblems, and trim. Squat down and check paint on exposed frame parts and fenderwells. Rubber bumper guards and vinyl trim should be dressed and lint-free. Carefully look into the

During the final inspection of the interior, wrap a piece of 2-inch masking tape around your hand with the sticky side out. Pat the cloth headliner, seats, and carpets to pick up lint and other light particles.

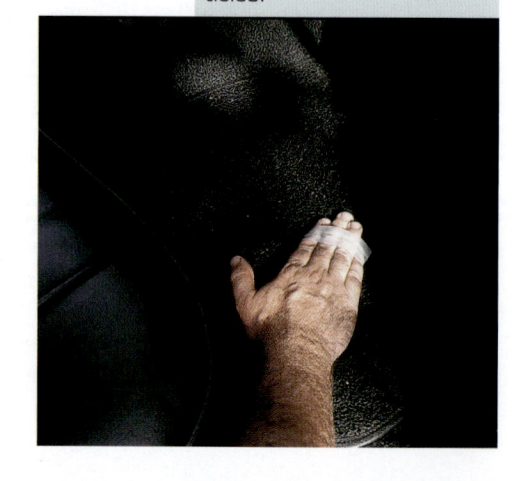

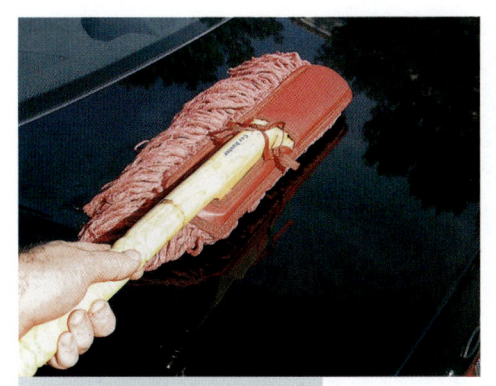

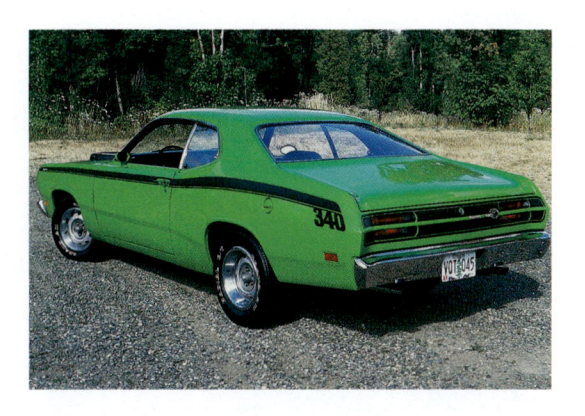

Above Left: The California Car Duster does a good job of removing dust from vehicle surfaces without worry of scratching paint. This may be a good option for those vehicles that are only driven during sunny weather and for which a regular wash is not feasible.

Above Right: In sunlight, you will easily detect smudges on glass and paint, dry polish or wax in nooks and crannies, spots on tires that were missed with dressing, and lots of other things that were not clearly visible under artificial light.

corners of the grille components and remove any remnants of dust and dirt. License plates clean? Plate frames look good?

Look at all of the glass. Sunlight will plainly show imperfections. Bull's eye damage to windshields must be repaired, or the full windshield may need to be replaced.

Look at the wheels and tires to determine if dressing was applied evenly. Many times the part of the tire that was on the ground at the time of initial dressing will not have been touched. Whitewalls should be checked for pattern and color uniformity. Tire cosmetologists can repair mismatched and damaged whitewalls; check with a local dealership or the Yellow Pages.

Chips on black lug nuts, black trim, and chassis can be touched up with a black felt pen. By no means a permanent repair, this quick fix will cover a blemish until you have the opportunity to properly sand, prime, and paint.

Dings in chrome bumpers can be cosmetically touched up with a tiny dab of bright silver paint. Spray a puddle on a piece of cardboard, dip a toothpick into it, and apply to the ding. From a couple of feet away, you won't know the difference.

The windshield wiper blades should be replaced if not working properly. If faded but in good shape, lightly buff the outer widths with just the dressing cloth; excess on the cloth will be sufficient, and you do not want to soak the wiper blades in dressing.

How do things look under the hood? In sunlight, you'll quickly notice paint flaws, dirt, and lint. Adjust ignition wires so they line up according to the design of the engine. Other wires should also run true to pattern. Remove any overspray with a dab of lacquer thinner on a rag. Chrome is touched up with chrome polish. Look at the underside of the hood for flaws.

Louvers are notorious places for wax buildup. Check them along the cowling at the base of the windshield. Use the cut-off paintbrush for cleaning.

Walk around the vehicle as many times as it takes. Look at it from every angle and height. Inspect every square inch. When you have made two trips around without finding a flaw, smile. Your work is done—good job!

Rattles and Squeaks

The longer you listen to a rattle or squeak, the more your ear gets used to it, finally tuning it out. Over time, the loose part will wear prematurely or break. On the other hand, as a true car enthusiast, seeking out new rattles and squeaks and repairing the cause will prolong the life and use of affected parts.

After a complete detail, go over the vehicle inside and out. Tighten screws, nuts, and bolts. Tighten trim, grille pieces, license plate frames, lug nuts, hood bolts, valve covers, seat brackets, dash screws, and so on. Use clean tools of the appropriate size and check every fastener you can see. As a preventive measure, you can do this once a month during the super cleaning or concentrate on just one section of the car with each washing. Use a system that ensures complete coverage once a month. During oil changes and lube jobs, concentrate on the screws, nuts, and bolts under the hood.

If you discover a rattle or squeak while driving, take note of it and plan to track it down later. Don't start looking for it while hanging onto the steering wheel doing 60 miles per hour down the highway. Let someone else drive the car while you hunt for the noise, pushing and pulling on parts as much as you please.

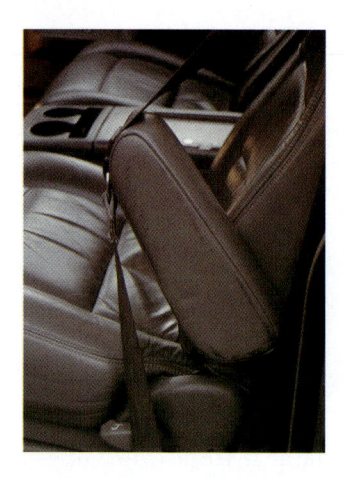

Far Left: During your final inspection, move things around to ensure you've done a thorough job of cleaning. When the armrest is pushed up in a stowed position, the seat belt rubs against it. These dark smudges must be wiped clean.

Left: This plastic hood deflector was washed and dried along with the rest of the vehicle. The side on the right was treated to Meguiar's Plastic Cleaner and Plastic Polish; the side on the left was not. The left side is dull and cloudy, while the polished side is dark and clear enough that a corner of the cloth is visible through it.

California Car Duster

A pristine automobile that sits idly parked in a garage will get dusty; there is not much you can do about it except put on a car cover. When dust appears, you can do one of two things: wash the entire vehicle or opt for use of the California Car Duster.

This tool is made of top-quality cotton treated with paraffin to attract surface dust. It is supposed to work better as it gets dirtier. Since it is made of cotton, you should not have to worry much about scratches, unless the vehicle is covered with dusty material that includes grit. Such might be drywall dust, sand, metal dust, and the like. You'll have to pay attention to what projects were completed in your garage to determine what the dust might be composed of.

After purchasing your California Car Duster, remove it from the bag, place it on a newspaper, and let it air out for about 48 hours. Remember to shake it vigorously before and after each use. Until this tool is "seasoned" after two or three uses, it may leave minor streaks behind from the paraffin wax. These streaks are easily and quickly wiped off with a clean towel.

For best service, lightly dust auto surfaces without applying pressure. The tool is a duster, not a scrubber. Use it on vehicles with cool surfaces and not on those that have just gone into the garage after sitting in the sun all day. This helps to protect against heating the paraffin wax, which could cause streaks.

The California Car Duster can be washed, if you determine it needs to be. Handwash in lukewarm water using a mild soap, like Woolite. Then allow it to air dry. Do not put it into a clothes dryer.

Quick Spray Spiff Products

When your vehicle is between washes, when it looks OK but you would like to quickly spiff it up, consider using a product like Meguiar's Quik Detailer or Mothers Showtime. Designed for use on vehicles with a good wax job, they will quickly remove surface dirt, dust, dulling contaminants, and fingerprints without scratching the finish. They are not designed to remove caked-on dirt.

Make sure your car's finish is cool to the touch; do not spray it on a vehicle that has been sitting in the sun. In the shade, hold the spray bottle about 2 feet from the surface and spray a small area. Use a soft clean cotton cloth to spread the material around the area. Flip over the cloth to a dry side to buff it off. Continue doing a small section at a time until the entire vehicle is clean and shiny.

Do not let the spray dry on the surface. If it does, simply spray more material on it and buff it off. If, after using the product a few times on different occasions, you notice that it is becoming more difficult to buff off, it is time to put a new coat of wax on the paint finish.

TECHNIQUE 59 PAINTLESS DENT REMOVAL

Time: 1/2 hour or more, depending on the size of the dent

Tools: Paintless Dent Repair Kit and a clean cloth

Talent: ★★★

Tab: $79–$299

Tip: Pull dents slowly

Gain: Quick and easy dent repairs

Complementary project: Wash and wax the vehicle

The Paintless Dent Repair Kit is manufactured by the Dent Fix Corporation and available through The Eastwood Company. It can be used on OEM baked paint finishes only. Kits vary in the amount of dent-pulling tools they contain and range in price from $79 to $299. Each kit contains a hot glue gun, glue sticks, a glue release agent, dent-pulling tools, and instructions. The key to the kits is the hot glue employed. One drop about the size of a garden pea is all that is needed. It is specially formulated and has been extensively tested to ensure that it forms a tight bond and will not damage factory OEM paint finishes.

After a dented area has been cleaned and gone over with a wax and grease remover, the dent and head of the pulling tool are cleaned with the release agent. Then a drop of hot glue from the glue gun is placed on the T-handled puller or stud. That tool is then placed at the center of the dent. Hold the tool in place for 10 to 15 seconds and allow the glue to cure for 2 to 3 minutes. Then slowly and firmly start pulling on the T-handle at a 90-degree angle or twisting on the wing nut to the stud and crossbar to pull the dent out.

Once the dent is pulled out, apply two drops of release agent on the glue to release the bond. The flat pad of glue will peel off easily, leaving the paint blemish-free.

1. A small dent is recognized by the deflection at the top of the white truck's chrome grille reflected in the paint from the background. Until the invention of paintless dent repair, a body shop would have had to repaint the spot after the dent was pulled out.

2. The dented area was cleaned, wiped off with a wax and grease remover, and cleaned with the Dent Out release agent. A pea-size ball of glue was placed on the tip of the threaded stud from the hot glue gun and the stud attached to the center of the dent.

Suction cups are used in attempts to pull out dents with some success. Most of the time, autobody shops have to grind away paint, weld studs to dents, and use special stud pullers to repair dents. The paintless system works very well and without the need to repaint.

T-handled pullers are used for a variety of dents, while the crossbar, tripod, and cone support tools offer a more solid base and a perfect 90-degree pull. The crossbar, tripod, and cone have a hole drilled in the center, which slips over a threaded stud. A wing nut is then screwed onto the stud and spun down until it makes contact with the support piece. The wing nut is slowly screwed tighter to put upward pressure on the stud and pull the dent out. Several attempts will likely have to be made on large dents.

The more you use this tool, the better you will become at dent repair. The process is quick and simple.

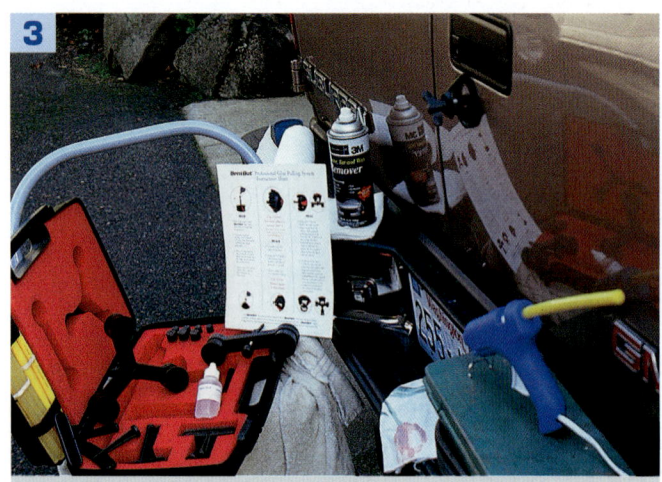

3. After waiting 2 to 3 minutes for the glue to cure, the cone support was placed over the stud and the wing nut was screwed onto the stud. The wing nut was slowly tightened down to put upward pressure on the stud to pull it and the dent out. The hot glue gun was placed on top of a metal toolbox as a safety precaution.

4. The reflection of the white truck's grille shows that the dent is gone, and the paint is in the same excellent condition it was in before the dent repair. The release agent caused the wad of glue to peel off easily.

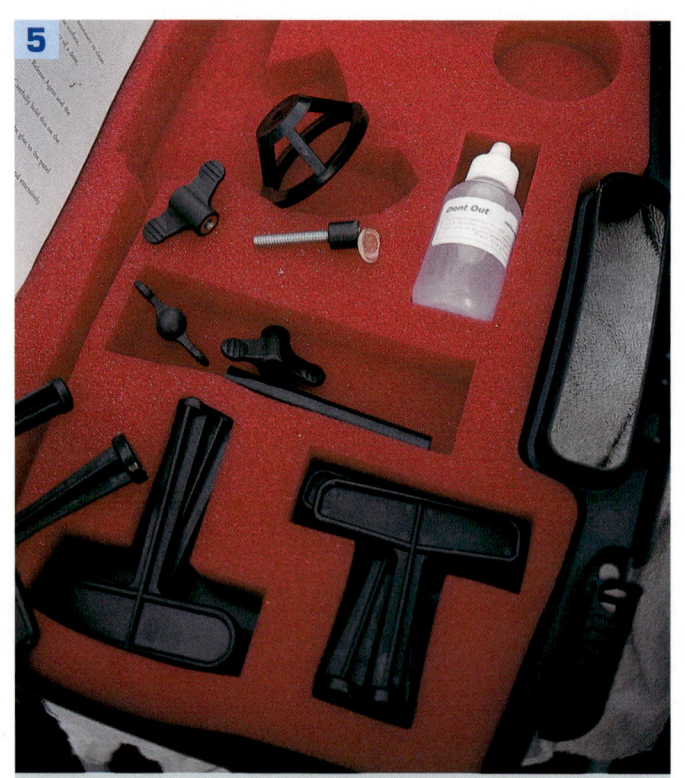

5. The wad of glue that was peeled off the dent repair in one piece lies next to the head of the stud used for the dent pull. This is the Professional Paintless Repair Kit, with a number of large and small T-handled pullers, large and small studs, a crossbar, tripod, and cone support.

Index

A
Air fresheners, 59, 148
All-purpose cleaners, 11–12
Antennas, 124
Ashtrays, cleaning, 58
Automatic car washes, 21

B
Battery box, detailed cleaning of, 71
Black trim, 117
Bra care, 149
Brightwork, preliminary washing, 26–27
Brushes, types and sizes of, 10–11
Buffing, with machine, 86–88
Bug splatter, removing, 25
Bumper stickers, removing, 32–33

C
California Car Duster, 156
Carburetor
  covering before detailing engine compartment, 63
  detailed cleaning, 69
Car covers, 149–150
Carnauba-based wax, 92
Carpet
  applying cloth protectants, 57
  shampooing, 12, 55–56
  vacuum cleaners for, 12–13
  vacuuming, 39–40
Chrome accessories
  polish and shine inside engine compartment, 75
  polishes for, 80
  polishing and waxing, 118
  trim cleaning, 116–117
Clay bar, 84–85
Cleaning solutions. *See also* Soaps
  all-purpose cleaners, 11–12
  chemical wheel cleaners, 133
  degreasers for initial engine detailing, 64
  for glass, 112–113
  for tires, 28
Cloth protectants, 57
Cloth seats
  dry cleaning, 51
  wet shampoo, 51–52
Concours d'Elegance, 151–153
  dismantling for preliminary washing, 17
  engine compartment tips, 153
  exterior tips, 152–153
  final thoughts about, 153
  interior tips, 151–152
  overview of, 151
  underbody tips, 153
Convertible tops, preliminary washing, 30–31

D
Dashboard
  cleaning, 43
  vinyl dressing for, 47–48
Decals
  in engine compartment, 77
  removing from glass, 111
  removing from paint, 32–33
  replacement of, 147
Degreasers, for initial engine detailing, 64
Dents, paintless dent removal, 157–158
Detailing
  degree of, 4
  generic, 4
  overview of, 4–5
Dismantling
  for Concours d'Elegance, 17
  for preliminary washing, 16–17
Distributor, covering before detailing engine compartment, 62–63
Doorjambs
  cleaning, 44–45
  preliminary washing, 22
Door panels, cleaning, 45–46
Drying
  importance of, 21
  preliminary washes, 21
Drying cloths, types of, 13

E
Emblems, in engine compartment, 77
Engine compartment, 60–77
  battery box detailing, 71
  caution for, 64–65
  cleaning each time you wash vehicle, 76–77
  clear lacquer, 76
  Concours d'Elegance tips, 153
  decals and emblems, 77
  detailed cleaning of engine, 69–70
  drying, 66
  fenderwell detailing, 70–71
  firewall area detailing, 70
  grille and radiator area detailing, 71
  hood underside, 65–66
  initial cleaning without water, 67–68
  initial cleaning with water, 64
  painting engine block and engine parts, 72–74
  polish and shine, 75
  preliminary wash at self-serve car wash, 9
  preparation for, 62–63
  removing paint overspray, 74
  work area, 62
Exterior detailing, 78–97
  buffing with a machine, 86–88
  clay bar, 84–85
  clear coats and urethane paint finishes, 89
  combating rust, 105
  Concours d'Elegance tips, 152–153
  final inspection for, 154–155
  hand polishing, 89–91
  hand waxing, 92–93
  oxidation damage, 81
  paint blemishes, 96
  paint chips, 96–97
  painting frame members, 106
  painting tailpipes, 106–107
  polish choices, 80
  removing polish and wax residue, 94–95
  removing severe oxidation, 82–83
  wax choices, 80–81
  ways to make exterior shine, 81

F
Fabric protectants, 57
Fender lip, 29
Fenderwell
  combating rust, 105
  detailed cleaning, 70–71
  painting, 73–74, 104–105
  polish and shine, 75
  thorough rinsing and cleaning, 102–103
  undercoating, 105
Finishing touches, 142–158
  air fresheners, 148
  bra care, 149
  California Car Duster, 156
  car covers, 149–150
  Concours d'Elegance, 151–153
  decal replacement, 147
  final inspection of exterior, 154–155
  final inspection of interior, 154
  paintless dent removal, 157–158
  part replacement, 146–147
  quick spray spiff products, 156
  rattles and squeaks, 155
  trunk cleaning, 143
Firewall area
  detailed cleaning of, 70
  painting, 73–74
  polish and shine, 75
Floorboards, repairing rusty, 38
Frame members, painting, 106

G
Glass detailing
  cleaning glass, 112–113
  cleaning window tint film, 113
  glass polishing, 114
  removing buildup from window edges, 115
  removing stickers and decals from glass, 111
  windshield repair, 121
Grille
  detailed cleaning of, 71
  preliminary washing, 26

H
Headers, detailed cleaning, 69
Hood
  cleaning underside, 65–66, 67
  detailed cleaning, 69
  preliminary washing, 23
Hubcaps, 138

I
ID badges detailing, 123
Interior detailing, 36–59
  air fresheners, 59
  cloth protectants, 57
  cloth seats, 51–52
  Concours d'Elegance tips, 151–152
  dashboard, 43
  doorjambs, 44–45
  door panels, 45–46
  final inspection for, 154
  interior dismantling, 38
  interior extras, 58–59
  leather seats, 53–54
  rear seat area, 41–42
  seatbelts, 43
  shampooing carpets, 55–56
  vacuuming, 39–40
  vinyl and rubber dressing, 47–48
  vinyl seats, 49–50

L
Leather seats, cleaning and conditioning, 53–54
License plate detailing, 123–124
Lug nuts, 139

M
Mag wheel cleaning, 136
Molding, cleaning and conditioning, 120

O
Oxidation
  damage from, 81
  removing severe, 82–83

P
Paint burn, 87
Paintless dent removal, 157–158
Paint/painting
  clear coats and urethane paint finishes, 89
  engine block and engine parts, 72–74
  fenderwells, 104–105
  firewall and inner fenders, 73–74
  frame members and tailpipes, 106–107
  polishes for, 80

removing paint overspray, 74
repairing paint chips and blemishes, 96-97
trunk, 144-145
waxing new paint, 93
wheels, 139
Part replacement, 146-147
Pinstripes, waxing, 93
Plastic light covers, 124
Plastic window care, 122
Polish
  applicators for, 89
  choosing, 80, 89-90
  engine compartment, 75
  function of, 80
  for glass, 114
  hand polish, 89-91
  for metal, 80
  for paint, 80
  removing residue, 94-95
  removing severe oxidation, 82-83
  sealer/glaze polish, 82-83, 89-90
Polishing compound, 80, 82-83
Preliminary washing, 15-35
  automatic car washes, 21
  brightwork, 26-27
  bug splatter, 25
  dismantling, 16-17
  doorjambs, 22
  drying, 21
  grille, 26
  hood edges, 23
  removing road tar, 25
  removing stickers and decals, 32-33
  removing tree sap, 24-25
  running boards, 27
  at self-serve car wash, 9
  steps in basic vehicle wash, 19-21
  tires, 28-29
  trunk edges, 23
  underbody, 34-35
  vinyl tops and convertibles, 30-31
  washing equipment for, 17-18
  wheels, 29
  work area for, 16
Preparation, 8-13
  appropriate attire, 8
  upholstery and carpet cleaning equipment, 12-13
  washing equipment, 9-13
  work area, 8-9
Pressure washers, 19
  damage from, 9
  for degreasing engine, 64-65

Q
Quick spray spiff products, 156

R
Racing stripes, waxing, 93
Radiator area, detailed cleaning of, 71
Rattles, fixing, 155
Rinsing, 19-21
  garden hose for, 9, 18
  underbody, 102-103
Road tar, removing, 25
Rubber
  cleaning and conditioning, 119
  dressing for interior detailing, 47-48
Rubbing compound, 82-83
Running boards, preliminary washing, 27
Rust, combating, 105
Rustproofing products, underbody, 101, 105

S
Sealer/glaze polish, 82-83, 89-90
Seatbelts, cleaning, 43
Seats
  cleaning front seat area, 42-43
  cleaning rear seat area, 41-42
  cloth, 51-52
  leather, 53-54
  vacuuming, 39-40
  vinyl, 49-50
Self-serve car wash, 9, 16
  for engine compartment, 62
Shampooing
  carpets, 55-56
  cloth seats, 51-52
Soaps. *See also* Cleaning solutions
  choosing, 9-10
  dish soap, 10, 17
  powdered, 10
  for preliminary wash, 17-18
  use of mild instead of chemicals, 4, 5
Spare tire, cleaning, 20-21
Squeaks, fixing, 155
Steam cleaning underbody, 101
Steel wool soap pads, 11
Stickers
  removing from glass, 111
  removing from paint, 32-33
  replacement of, 147

T
Tailpipes, painting, 106-107
Tires
  cleaning supplies for, 28
  cleaning whitewalls, 28-29, 129-130
  preliminary washing, 28-29
  raised white lettering, 129-130
  tire dressing and tire black, 131-132
Tree sap, removing, 24-25
Trim detailing
  black trim, 117
  chrome polishing and waxing, 118
  cleaning, 116-117
  rubber and vinyl cleaning and conditioning, 119
Trunk
  cleaning, 143
  preliminary washing, 23
  rejuvenation, 144-145
Turbochargers, caution for, 65

U
Underbody detailing, 98-107
  Concours d'Elegance tips, 153
  fenderwell painting and undercoating, 104-105
  preliminary washing, 34-35
  rustproofing, 101
  steam cleaning, 101
  thorough rinsing and cleaning, 102-103
  undercoating, 101
Undercoating
  fenderwells, 104-105
  function of, 101
Upholstery
  applying cloth protectants, 57
  cleaning cloth seats, 51-52
  drying cloths for cleaning, 13
  repairing tears, 58
  shampoos for, 12
  stain removal, 51
  vacuum cleaners for, 12-13
  vacuuming, 39-40

V
Vacuuming, for interior detailing, 39-40
Vinyl
  bra care, 149
  cleaning and conditioning trim, 119
Vinyl dressing, 42
  for interior detailing, 47-48
  for vinyl seats, 50
Vinyl seats
  cleaning, 49-50
  dressing, 50
  seat repair, 50
Vinyl tops, preliminary washing, 30-31

W
Washing equipment
  all-purpose cleaners, 11-12
  brushes, 10-11
  car wash soaps, 9-10
  for preliminary washing, 17-18
  steel wool soap pads, 11
  wash mitts, 10
  water applicators, 9
Wash mitt, 19
  synthetic vs. cotton, 10
  using, 17
Wash wands, 9
Water applicators, 9
  preliminary washing, 18
Wax
  carnauba-based, 81, 92
  choosing, 80-81, 92-93
  cloth for application, 13
  cloth for removing, 13
  function of, 80
  hand waxing, 92-93
  new paint, 93
  one-step cleaner waxes, 80-81
  pinstripes and racing stripes, 93
  protection from oxidation by, 81
  removing residue, 94-95
  times per year recommendation, 92
Wheel covers, 138
Wheels
  chemical wheel cleaners, 133
  cleaning other special wheels, 136-137
  lug nuts, 139
  mag cleaning, 136
  painting, 139
  preliminary washing, 29
  removing for preliminary washing, 16-17
  wheel covers and hubcaps, 138
  wire wheel cleaning, 134-135
Whitewalls
  cleaning, 129-130
  preliminary washing, 28-29
Windows. *See also* Glass detailing
  plastic window care, 122
  window tint film, 113
  windshield repair, 121
Windshield wipers
  cleaning and conditioning, 125
  washing, 20
Wire wheel cleaning, 134-135
Work area, 8-9
  engine compartment, 62
  for preliminary wash, 8-9, 16